Calculus Applications in Engineering and Science

Stuart Goldenberg and Harvey Greenwald
California Polytechnic State University

keyed to

Calculus
Fourth Edition
Larson/Hostetler/Edwards

D. C. Heath and Company
Lexington, Massachusetts   Toronto

International Standard Book Number: 0-669-21676-3

2 3 4 5 6 7 8 9 0

PREFACE

Calculus Applications in Engineering and Science contains applications of current probems in engineering, physics, chemistry, biology and related fields. These applications are developed with full numerical data — much of it drawn from real situations — to provide a context for student practice with calculus theory and techniques. The applications in this book can be used in a calculus or physics course in a number of ways, including as a source of problems for lecture topics, group projects, individual or class assignments, and test items. Each application in this book begins with a detailed description of the topic, followed by an example. These examples include carefully detailed, step-by-step solutions. The exercises that follow give students the opportunity to use their calculus skills to solve new problems related to the applications.

Calculus Applications in Engineering and Science addresses the primary topics of calculus and is intended for engineering, physics, and science majors. Each application is keyed by section number and title to the corresponding material in *Calculus, Fourth Edition* by Larson/Hostetler/Edwards. This feature allows the instructor to integrate additional applications into the course easily, and it facilitates student self-study. For students and instructors using a different textbook, application titles clearly indicate the topics covered.

The authors wish to thank the staff of D. C. Heath and Company for their assistance in the preparation of this book: Ann Marie Jones, Acquisitions Editor; Cathy Cantin, Developmental Editor, and Carolyn Johnson, Editorial Associate. To these people and others who have helped us, we express our appreciation.

Stuart Goldenberg
Harvey Greenwald

CONTENTS

1

THE CARTESIAN PLANE AND FUNCTIONS

APPLICATION 1.1: Sound and Light Distribution

> See Section 1.2
> The Cartesian Plane
> Calculus, 4th Edition,
> Larson/Hostetler/Edwards

A large room contains two sound speakers that are D feet apart; one speaker is twice as loud as the other. (See Figure 1.1.)

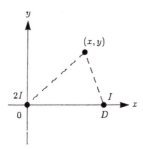

FIGURE 1.1

Suppose the speakers themselves cannot be adjusted, but the listener is free to move about to find those positions that give equal amounts of sound from both speakers. The desired location satisfies two conditions: sound at the listener's position is (1) directly proportional to the sound level of a source and (2) inversely proportional to the square of the distance from the source.

Similar relationships hold true for light sources. If walls are present, their reflective properties also play a role.

EXAMPLE

Sound Speakers

Given the speakers in Figure 1.1, find the equation describing all places where one could stand to have equal amounts of sound from both sources.

SOLUTION

See Figure 1.1. Let $I =$ the intensity of the softer speaker. Then $2I$ is the intensity of the louder one. We equate the amount of sound from each speaker.

$$\frac{2I}{x^2 + y^2} = \frac{I}{(D - x)^2 + y^2}$$

$$2(D - x)^2 + 2y^2 = x^2 + y^2$$

$$x^2 - 4xD + 2D^2 + y^2 = 0$$

$$(x - 2D)^2 + y^2 = 2D^2$$

This is a circle of radius $D\sqrt{2}$ and center at $(2D,\ 0)$, as shown in Figure 1.2.

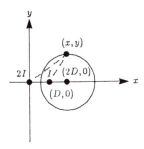

FIGURE 1.2

APPLICATION 1.1 EXERCISES

1. Two loudspeakers are 20 feet apart. One speaker is 70% as loud as the other. Find an equation describing all points where a person could stand to have equal amounts of sound from both sources.

2. Two light sources are 100 meters apart. One light is 10 times as bright as the other. Find all points at which the amount of light will be the same from both sources.

3. Two light sources are 10 meters apart. One light is 90% as bright as the other. Find all points at which the amount of light will be the same from both sources.

4. The mass of the earth is 5.98×10^{24} kilograms and the mass of the moon is 7.36×10^{22} kilograms. Assuming that the positions of these bodies are fixed, find all points where the gravitational attraction toward the moon and toward the earth will be the same. The earth and the moon are 3.84×10^{8} meters apart. [*Hint:* Newton's Universal Law of Gravitation is

$$F = \frac{Gm_1 m_2}{r^2}$$

where F is the attractive force between the objects, G is a constant, m_1 and m_2 are the masses, and r is the distance between the centers of the two masses.]

APPLICATION 1.2: Tree Growth

> See Section 1.3
> **Graphs of Equations**
> **Calculus, 4th Edition,**
> **Larson/Hostetler/Edwards**

A giant Sequoia tree, perhaps the largest tree in the world, stands 272 feet tall and is about 34 feet in diameter at its base. Its habitat is the central California mountain range, the Sierra Nevada, in the Sequoia National Park. The great age, size, and rapid growth of this tree have contributed to its fame. It has the heroic name *General Sherman*.

EXAMPLE

Calculating Volume

We can approximate the volume of a tree by disregarding the branch structure and considering the trunk as a cylinder. If the General Sherman has an average radius of 7.6 feet, what is its volume?

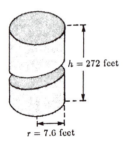

FIGURE 1.3

SOLUTION

Assuming the average radius of the tree is 7.6 feet, we approximate the tree as a long cylinder. (See Figure 1.3.) The volume of a cylinder is $\pi r^2 h$. Therefore, we have

$$V = \pi r^2 h = \pi (7.6^2) 272 \approx 4.94 \times 10^4 \text{ ft}^3.$$

APPLICATION 1.2 EXERCISES

1. Assuming the General Sherman produces an annual growth ring of 0.003 feet, how much new wood is added to the tree (trunk) each year?

2. Suppose the paper industry cultivated a fast-growing tree that could grow from a seedling to a height of 50 feet in one year. If the increase in volume of this tree were to match that of the General Sherman (see Exercise 1), what would be the average radius of the tree?

3. The baobab tree, native to Australia and Africa, has a trunk that measures as much as 60 feet in diameter. However, this tree only grows to a height of 40 feet. Approximate the amount of new wood this tree would produce each year. Use a cylindrical model with a radius of 30 feet, a height of 40 feet, and an annual growth ring of 0.003 feet.

4. Model the baobab tree described in Exercise 3 as a cone. Approximate the amount of new wood it would produce in one year if the increase in thickness is 0.003 feet.

APPLICATION 1.3: Expanding Gases

> See Section 1.3
> Graphs of Equations
> Calculus, 4th Edition,
> Larson/Hostetler/Edwards

Perfect gases expand when heated according to the Ideal Gas Law

$$PV = nRT$$

where P is the pressure, V is the volume, n is proportional to the amount of gas present, R is a constant, and T is the Kelvin temperature. Engineers who design cars use this formula to determine the "explosion point," that is, the condition when increase in temperature will cause the windshield of a car to pop out.

EXAMPLE

Closed Car

Assume the interior of a locked car is airtight, the contents are at one atmosphere, and the initial temperature is 293 Kelvin (about 68° F). Suppose the temperature inside the car increases 15 Kelvin for each hour the car remains in the sun. The car manufacturer has designed the windshield so that it will pop out when the interior pressure is 1.6 times the exterior pressure. How long would these conditions have to be maintained before the windshield pops out?

SOLUTION

Using the Ideal Gas Law, compare the interior of the heated car to the initial condition of the car.

$$\frac{P}{P_0} = 1.6$$

$$\frac{(nRT)/V}{(nRT_0)/V} = 1.6$$

$$T = 1.6T_0$$

$$293 + 15t = (1.6)(293)$$

$$15t = (0.6)(293)$$

$$t \approx 11.72 \approx 12 \text{ hours}$$

APPLICATION 1.3 EXERCISES

1. A car with a sun roof will heat up faster than a car with a solid roof. Suppose the temperature of the interior of a closed car with a sun roof increases at a rate of 20 Kelvin per hour. Suppose the same pressure difference as described in the example will cause the windshield to pop out. How long will it take to reach this pressure?

2. Older cars do not have windshields that pop out. If an older car has a sun roof and the interior temperature increases by 20 Kelvin per hour, what will the interior temperature be after $1\frac{1}{2}$ hours in the sun? (Assume the initial temperature is 293 Kelvin.)

3. A half-full bottle of sunscreen is left in a locked car. One window is partially open so the air pressure in the car does not build up, but the temperature does. The temperature in the bottle increases from 20 degrees Celsius to 80 degrees Celsius. Find the pressure within the bottle of sunscreen. Assume there is no change in volume and there is no increase in vapor pressure from the heated lotion.

4. The plastic bottle described in Exercise 3 is empty and it expands. The volume is 15 percent more when heated. What is the new pressure?

APPLICATION 1.4: Temperature Scales

> See Section 1.4
> Lines in the Plane
> Calculus, 4th Edition,
> Larson/Hostetler/Edwards

The 1987 Nobel Prize in Physics was awarded to two researchers, Drs. Bednorz and Müller, of IBM's Zurich Research Lab for their discoveries in high-temperature superconductivity. Superconductivity refers to the fact that, at low enough temperatures, many substances lose all resistance to the passage of electricity. This phenomenon had long been known, but because the temperature at which substances displayed superconductivity was so low, practical applications were limited. Müller and Bednorz succeeded in finding a ceramic compound that lost resistance to electrical current at a surprisingly high "low temperature." Due to this breakthrough discovery, we now have materials that will be superconductors above the temperature of 77 Kelvin, which is the liquid/gas phase transition of nitrogen.

Temperature can be described on various scales: Kelvin, Fahrenheit, and Celsius. We know that water boils at 212 degrees Fahrenheit (373 Kelvin or 100 degrees Celsius), and freezes at 32 degrees Fahrenheit (273 Kelvin or 0 degrees Celsius). Using these facts, we can convert temperature expressed in one system to another system.

EXAMPLE

Superconductivity

Superconductivity is shown at 77 Kelvin. What is the corresponding Fahrenheit temperature?

SOLUTION

We begin by obtaining the equation which will convert a Kelvin temperature to a Fahrenheit temperature. Using the known boiling and freezing points of water, we have the following.

$$F = a\text{K} + b$$
$$212 = 373a + b$$
$$32 = 273a + b$$
$$180 = 100a$$
$$a = \frac{9}{5} \quad \text{and} \quad b = -459.4$$

We now have the conversion formula.

$$F = \frac{9\text{K}}{5} - 459.4$$

Substituting for 77 Kelvin, we obtain

$$F = \frac{9(77)}{5} - 459.4 = -320.8° \text{ Fahrenheit.}$$

APPLICATION 1.4 EXERCISES

1. What is the conversion equation relating a given temperature in Fahrenheit to one in Kelvin?

2. An automobile mechanic is asked to flush a radiator and add the correct amount of coolant and water to provide protection from freezing to -40 degrees. The customer fails to tell the mechanic if he means degrees Celsius or degrees Fahrenheit. The mechanic completes the job correctly without this information. Explain how he was able to complete the job to the customer's satisfaction.

3. The temperature of the surface of the sun is 5525 degrees Celsius. Find the equivalent temperature in both the Kelvin and Fahrenheit scales.

4. Sun spots are 1500 degrees Celsius cooler than the rest of the surface of the sun. Find this temperature difference in the Fahrenheit scale.

APPLICATION 1.5: Potential Differences Between Capacitors

> See Section 1.4
> Lines in the Plane
> Calculus, 4th Edition,
> Larson/Hostetler/Edwards

A parallel-plate capacitor is an electric circuit element used to store charge. (See Figure 1.4.) Between the plates of a parallel plate capacitor the electric field is constant. If we are given initial voltages on each plate and the distance between the plates, we can write a linear equation to describe the potential between the plates. From this equation, we can then see that varying the distance between the plates will affect the voltage between them.

EXAMPLE

Calculating Voltage in a Capacitor

As shown in Figure 1.4, the voltage on one of two parallel plates is -50 volts and on the other, $+100$ volts. The plates are 5mm apart. Determine a linear equation that describes the voltage in the space between the plates as a function of the distance between the plates.

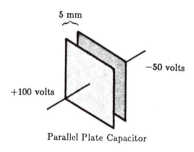

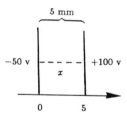

FIGURE 1.4

SOLUTION

Using a linear model, the voltage between the parallel plates can be written as

$$V = ax + b$$

where x is the distance in millimeters from the plate that has a voltage of -50. Substituting, we have

$$V(0) = -50 = b.$$

Hence, $b = -50$. Moreover, since $V(5) = 100$, we have

$$100 = 5a - 50$$
$$30 = a.$$

Thus, the linear model is

$$V = 30x - 50, \quad 0 \le x \le 5.$$

APPLICATION 1.5 EXERCISES

1. The relationship between the electric field, E, and the voltage in a parallel-plate capacitor is $V = Ed$, where d is the distance between the two plates. Find the electric field for the parallel-plate capacitor described in the example.

2. The dielectric strength of air is 3×10^6 volts/meter. If an electric field is greater than this, the air breaks down and electricity will flow through the air. For the parallel plates in our example, the voltage between the two plates is maintained at 150 volts. How close together can the plates be before the air between the plates breaks down?

3. At the surface of a spherical conductor, the relationship between the voltage and the electric field is $V = RE$, where V is the voltage or potential, E is the electric field, and R is the radius of the conductor. If the dielectric strength of air is 3×10^6 volts/meter, what is the breakdown voltage for a spherical conductor of radius 0.005 meters?

4. The spherical conductor described in Exercise 3 is placed in a plastic bath of dielectric constant 4. The result is a dielectric strength of 1.2×10^7. What is the breakdown voltage for this situation?

APPLICATION 1.6: Water Rationing

> See Section 1.5
> Functions
> Calculus, 4th Edition,
> Larson/Hostetler/Edwards

The city of San Luis Obispo, California, has a water conservation system, initiated in 1989. In order to ration water, users are encouraged to restrict their water usage and are also assessed penalties for overuse.

Under the conservation system, each family has a water allocation, which is 70% of their 1987 summer months' usage. If the family uses less than this amount, the water charge is $1.50 per water unit. (One water unit is 100 cubic feet.) If the family uses more than their allocation, but less than the entire amount used during the 1987 summer months' usage, the rate is double the regular rate. If the family uses more than the entire amount used during the summer of 1987, the rate is triple the regular rate. In each case the rate applies to *all* the water units used. Thus, if a family uses more than they did in 1987, they will be charged at the maximum rate for all water used. However, if they restrict their usage to less than 70%, they will have all their water units charged at the minimum rate.

EXAMPLE

Calculating Water Usage

As an example of how the San Luis Obispo program works for a billing cycle, we shall consider a typical family, the Morris family, that used 45 water units per billing cycle in 1987. Describe the Morris family's 1989 water allocation using the San Luis Obispo rate schedule.

SOLUTION

Since the 1989 allocation is 70% of their 1987 usage, which is 45 units per billing cycle, the Morris family would be allocated 32 water units per billing cycle. (The water department rounds 31.5 to 32.) If x units are used, and x is between 0 and 32 units, the charge is $1.50 per unit, or $1.5x$. If usage exceeds 32 units but is still less than the 1987 usage of 45 units, then each unit costs $3.00, or $3x$. The maximum cost occurs for usage in excess of the 1987 usage. The maximum rate is three times the usual rate, or $4.50 per unit. The fee for water would then be $4.5x$ where, once again, x is the total number of water units used. Using function notation, we can write all three rates as

$$p(x) = \begin{cases} 1.5x, & 0 \le x \le 32 \\ 3x, & 32 < x \le 45 \\ 4.5x, & 45 < x \end{cases}$$

where $p(x)$ is the cost of x units of water.

APPLICATION 1.6 EXERCISES

1. Construct a function that shows the charges for water in 1989 if the Morris family's usage is between 10 and 50 units. Sketch a graph of this function.

2. Another possible model for water rationing is to increase rates for only those water units that exceed a certain amount. In other words, a family would be charged $3.00 for usage in excess of 70% for each additional unit until the 1987 level in reached. Only that usage in excess of the 100% level would be charged at the rate of $4.50 per unit. Construct a function for this rate schedule.

3. The drought of 1989 has affected other cities of the central California coast. To promote water rationing, Monterey has a rate schedule structured with step-by-step rate increases. Monterey households are limited to 80% of their 1987 water usage. Water used within the allotment has a cost of $1.50 per unit. The first two units used over the allotment are charged at $25 each. The next two units are $50 each, and so on until the charges reach a maximum of $250 per unit. Using the Monterey plan, construct a function that shows the charges for water if the usage in 1987 was 45 units and the 1989 usage is between 10 and 50 units.

4. During the winter months in San Luis Obispo, water rationing is based on 90% of the 1987 usage rather than 70%. All other conditions are the same. This change in allocations is reasonable because in the winter, water demand is much lower. Suppose the Morris family used 30 water units in a winter billing period in 1987. Construct a function that shows the charges for usage between 12 and 35 units.

2

LIMITS AND THEIR PROPERTIES

APPLICATION 2.1   Escape Velocity
APPLICATION 2.2   Resonance

APPLICATION 2.1: Escape Velocity

> **See Section 2.5**
> **Infinite Limits**
> **Calculus, 4th Edition,**
> **Larson/Hostetler/Edwards**

In order to escape the earth's gravitational pull, a rocket must be launched with a certain initial velocity. This velocity is called the escape velocity. A rocket launched from the surface of the earth (approximate radius: 4000 miles) has velocity, v, given by

$$v = \sqrt{\frac{2GM}{r} + v_0{}^2 - \frac{2GM}{R}}$$

$$\approx \sqrt{\frac{192,000}{r} + v_0{}^2 - 48} \text{ mi/sec}$$

where v_0 represents the initial velocity, r represents the distance from the rocket to the center of the earth, G is the gravitational constant, M is the mass of the earth, and R is the radius of the earth.

EXAMPLE

Calculating Escape Velocity

Find the value of v_0 for which we obtain an infinite limit for r as v approaches 0. This value for v_0 is the escape velocity for earth.

SOLUTION

We first solve the given equation for r.

$$v^2 = \frac{192,000}{r} + v_0{}^2 - 48$$

$$\frac{192,000}{r} = v^2 - v_0{}^2 + 48$$

$$r = \frac{192,000}{v^2 - v_0{}^2 + 48}$$

We next evaluate the limit as $v \to 0$.

$$\lim_{v \to 0} \frac{192,000}{v^2 - v_0{}^2 + 48} = \frac{192,000}{48 - v_0{}^2}$$

This limit is infinite if $v_0{}^2 = 48$ or $v_0 = \sqrt{48}$ mi/sec. The value $\sqrt{48}$ mi/sec is the escape velocity for earth. If $v_0 < \sqrt{48}$, this limit represents the maximum distance away from the earth that the object will travel.

APPLICATION 2.1 EXERCISES

1. If $v_0 > \sqrt{48}$, show that the velocity v is always positive. In this case, the rocket will travel indefinitely.

2. A rocket launched from the surface of the moon has velocity given by

$$v = \sqrt{\frac{1920}{r} + v_0{}^2 - 2.17} \text{ mi/sec.}$$

Find the escape velocity for the moon.

3. A rocket launched from the surface of Mars has velocity given by

$$v = \sqrt{\frac{20,544}{r} + v_0{}^2 - 9.55}\ \text{mi/sec.}$$

Find the escape velocity for Mars.

4. A rocket launched from the surface of Planet X has velocity given by

$$v = \sqrt{\frac{10,600}{r} + v_0{}^2 - 6.99}\ \text{mi/sec.}$$

Find the escape velocity. Is this planet larger or smaller than earth? (Assume that the mean density of Planet X is the same as that of the earth.)

APPLICATION 2.2: Resonance

> See Section 2.5
> Infinite Limits
> Calculus, 4th Edition,
> Larson/Hostetler/Edwards

When a spring is displaced from its equilibrium position and released, it will oscillate with a frequency called the natural frequency. Resonance occurs when an external force is *in phase* with this oscillation, meaning that the external force is always acting in the direction of motion.

Resonance can be either destructive or constructive. Destructive resonance has resulted in the collapse of a number of bridges. For example, destructive resonance caused by a severe windstorm resulted in the collapse of the Tacoma Narrows Bridge in 1940. Constructive resonance, on the other hand, is used to advantage in the tuning of radios.

EXAMPLE

Determining Resonant Frequency

A spring with an attached object of mass m and spring constant k is acted upon by an external force $F(t) = C\cos(\omega t)$ where C is a constant. If we ignore resistance, the position at time t is given by

$$x = C_1 \cos\left(\sqrt{\frac{k}{m}}\,t\right) + C_2 \sin\left(\sqrt{\frac{k}{m}}\,t\right) + \frac{C\cos(\omega t)}{(k/m) - \omega^2}.$$

where C_1 and C_2 are constants and x represents the distance from the equilibrium position. Determine the angular frequency, ω, for which resonance will occur in a spring. [Note that the frequency f is related to the angular frequency ω by the equation $\omega = 2\pi f$.]

SOLUTION

We wish to determine the value of ω for which x will have an infinite limit. We set $(k/m) - \omega^2 = 0$ to obtain $\omega = \sqrt{k/m}$. We know that

$$\lim_{\omega \to \sqrt{k/m}} \left[C_1 \cos\sqrt{\frac{k}{m}}\,t + C_2 \sin\sqrt{\frac{k}{m}}\,t + \frac{C\cos\omega t}{(k/m) - \omega^2} \right]$$

is unbounded. For frequencies close to $\sqrt{k/m}$, the amplitude of the oscillations becomes large. This situation in which we assume no resistance is often referred to as pure resonance. [**Note:** The spring constant k is given in lb/ft.]

APPLICATION 2.2 EXERCISES

1. A 16-pound weight is attached to a spring and stretches the spring 8 inches. (See figure.) After the spring reaches its equilibrium position, an external force given by $F(t) = C\cos(\omega t)$ acts upon the spring. Determine the frequency for which resonance will occur in the spring. [*Hint:* Mass and weight are related by the equation $F = ma$, where $a \approx 32$ ft/sec² is the acceleration due to gravity.]

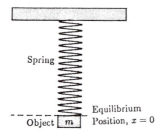

FIGURE FOR EXERCISE 1

2. A 4-pound weight is attached to a spring. If an external force given by

    $$F(t) = 10\cos 2t$$

    results in resonance, determine the spring constant k.

3. An object of unknown mass is attached to a spring with spring constant $k = 20$ pounds/feet. If an external force given by

    $$F(t) = 20\cos 4t$$

    results in resonance, determine the mass of the object.

4. The position of a spring at time t is given by

    $$x = 10\cos 2t + 10\sin 2t + \frac{100\cos \omega t}{4 - \omega^2}.$$

    Determine the frequency, ω, for which resonance will occur in the spring.

3

DIFFERENTIATION

APPLICATION 3.1: Radio Reception

> See Section **3.1**
> **The Derivative and the Tangent Line Problem**
> **Calculus, 4th Edition,**
> **Larson/Hostetler/Edwards**

In hilly areas, reception for both television and radio is frequently poor. We consider a situation where an FM transmitter is located behind a hill, and a radio receiver is on the opposite side of the hill. (See Figure 3.1.) How far from the base of the hill should a radio be located so that its reception is not obstructed? Obviously, the height of the transmitter, compared with the height of the hill is crucial, as is the specific positioning of the base of the transmitter. The radio can receive a clear signal if located far from the hill, provided the signal is strong enough. What is the *closest* the radio can be to the hill so that reception is not obstructed?

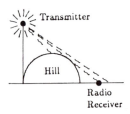

FIGURE 3.1

EXAMPLE

Tangent Lines

We consider an idealized situation: the contour of the hill is a semicircle of radius 1 unit. The transmitter is at the base of the hill and has a height of 2 units. The radio is on the side of the hill opposite the transmitter.

Figure 3.2 illustrates the situation. The position of the *tangent line TR* is the key element, where T is the top of the transmitter and R is the radio receiver. P is the point where the tangent line intersects the circle. We wish to find R. In other words, how far from the base of the hill should the radio be placed in order for it to receive an unobstructed signal from the transmitter?

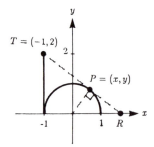

FIGURE 3.2

SOLUTION

In Figure 3.2, find the point R, where the tangent line to the circle through the point $(-1, 2)$ crosses the x-axis. We do this in three steps. First we form a general equation describing a line from the tower T to the receiver R, tangent to the hill. Then we determine the coordinates, (x, y), of the specific point P where TR meets the hill. Finally, by substitution, we will have the particular equation for tangent TR and we will be able to pinpoint the location of R.

The slope of the radius line at the point (x, y) is y/x. The tangent line is perpendicular to the radius line and thus has slope $-x/y$. Hence, points on a tangent line satisfy the equation

$$\frac{y-2}{x+1} = -\frac{x}{y}$$
$$x^2 + y^2 = 2y - x.$$

The equation of the circle is given by $x^2 + y^2 = 1$. We solve the equation of the circle and of the tangent simultaneously to obtain

$$2y - x = 1$$
$$x = 2y - 1.$$

Substituting this result into the equation of the circle, we obtain

$$4y^2 - 4y + 1 + y^2 = 1$$
$$5y^2 - 4y = 0$$
$$y(5y - 4) = 0.$$

Hence, $y = \frac{4}{5}$ and so $x = 2\left(\frac{4}{5}\right) - 1 = \frac{3}{5}$. This gives us the coordinates of P as $\left(\frac{3}{5}, \frac{4}{5}\right)$. (Note that the solution $y = 0$, $x = -1$ is not a solution to our problem because it lies on the wrong side of the hill.) The equation of the tangent line then becomes

$$\frac{y-2}{x+1} = \frac{-3/5}{4/5} = -\frac{3}{4}$$
$$3x + 4y = 5.$$

We can now pinpoint the location of R. The line $3x + 4y = 5$ has as its x-intercept $x = \frac{5}{3}$. The radio must be placed at least $\frac{2}{3}$ unit from the base of the hill.

APPLICATION 3.1 EXERCISES

1. Solve the problem as given in the example if the transmitting tower (of height one unit) is placed on the top of the hill.
2. Consider the problem as given in the example, but suppose the height of the transmitter is unknown. If it is known that the radio must be placed at least one unit from the base of the hill in order to receive an unobstructed signal, determine the height of the transmitter.
3. Solve the problem as given in the example if the hill contour is described by the equation $y = x - x^2$.
4. Solve the problem as given in the example, but suppose the transmitter is located at the point $(-1, 4)$.

APPLICATION 3.2: Bouncing Ball

> **See Section 3.2**
> **Velocity, Acceleration, and Other Rates of Change**
> **Calculus, 4th Edition,**
> **Larson/Hostetler/Edwards**

In sports such as racquetball and tennis, the player instinctively uses the rebounding of the ball to advantage. In manufacturing processes where an item is dropped from one piece of equipment into another, or into packaging, the amount of "bounce" must be considered.

Our situation here deals with the idealized ball, dropped vertically, from a known height, with a known rebound. We can determine a function for the ball's velocity from the time of the first bounce up to the second bounce. To solve this problem, we will use the formula for determining the position of a falling body

$$s = -\frac{1}{2}gt^2 + v_0 t + s_0$$

where g is the acceleration due to gravity, v_0 is the initial velocity, and s_0 is the initial position. We will also use the formula for determining the velocity of a falling body

$$v = \frac{ds}{dt} = -gt + v_0$$

where $g = 32$ ft/sec^2.

EXAMPLE

Rebound Velocity

A tennis ball dropped from a height of 16 feet rebounds to a height of 4 feet.

(a) Find the formulas for the velocity $v(t)$ valid for any time t before the ball lands a second time.

(b) Evaluate $\lim\limits_{t \to 1^+} v(t)$ and $\lim\limits_{t \to 1^-} v(t)$.

(c) In this model, is velocity a continuous function of time?

SOLUTION

Let s represent the position of the tennis ball at any time t.

(a) We determine the velocity in two steps; first for the period up to the first bounce, then for the period after the first bounce but prior to the second. Since

$$s = 16 - 16t^2$$

the ball lands the first time when $s = 0$ and $t = 1$. So,

$$v = \frac{ds}{dt} = -32t \quad \text{for } 0 \leq t < 1.$$

We next deal with the time period prior to the second landing of the ball. Let $1 < t$ and let v_0 represent the initial velocity in this time period. Then

$$s = v_0(t - 1) - 16(t - 1)^2$$

and so

$$v = \frac{ds}{dt} = v_0 - 32(t - 1).$$

At the maximum height, $s = 4$ and $v = 0$.
We solve simultaneous equations for v_0 and t.

$$4 = v_0(t - 1) - 16(t - 1)^2$$

$$0 = v_0 - 32(t - 1)$$

Rewriting the second equation gives

$$v_0 = 32(t - 1)$$

and so

$$(t - 1) = \frac{v_0}{32}.$$

Thus, substituting for $(t - 1)$ yields

$$\frac{v_0{}^2}{32} - 16\left(\frac{v_0}{32}\right)^2 = 4.$$

Hence, $v_0 = 16$ and $v = 16 - 32(t - 1)$. Since $(t - 1) = v_0/32$, $t - 1 = 1/2$ second, and hence, the total time in this time period is $2(1/2) = 1$ second. Thus,

$$v = \begin{cases} -32t, & 0 \le t < 1 \\ 16 - 32(t - 1), & 1 < t < 2. \end{cases}$$

We see that velocity thus has two functional formulas, depending on the value of t.

(b) $\lim\limits_{t \to 1+} v(t) = \lim\limits_{t \to 1+} [16 - 32(t - 1)] = 16$ ft/sec

$\lim\limits_{t \to 1-} v(t) = \lim\limits_{t \to 1-} [-32t] = -32$ ft/sec

(c) No, v is not continuous because the limits given in (b) are not equal.

APPLICATION 3.2 EXERCISES

1. Consider the problem as given in the example, except suppose the height of the rebound is unknown. (a) If the rebound velocity is 8 feet/second, what is the height of the rebound? (b) Find the formula for the velocity for any time t before the ball lands a second time.

2. Consider the problem as given in the example, except suppose the height of the rebound is unknown. If the total time from the point at which the ball is dropped until it lands a second time is 1.25 seconds, what is the formula for the velocity, valid for any time t before the ball lands a second time?

3. Solve the problem as given in the example, except assume that the ball rebounds to a height of 12 feet.

4. Consider the problem as given in the example, except assume that the ball rebounds a second time to a height of 1 foot. Find the formula for the velocity $v(t)$ that is valid for any time t before the ball lands a third time.

APPLICATION 3.3: Chain Rule

> **See Section 3.5**
> **The Chain Rule**
> **Calculus, 4th Edition,**
> **Larson/Hostetler/Edwards**

By now, pictures of astronauts and their equipment floating with their space vehicle are familiar. Here the concepts of weight and weightlessness are quite visibly demonstrated. We know intuitively now that an object at a distance from the earth has less weight than if measured on the earth's surface. But what of the *rate of change* of weight, as an object rises?

Let's consider a rocket carrying a satellite of known weight on earth. To determine the rate of change of the weight of an object, we need to know the force, F, between the earth and the satellite as a function of the distance, r, between them. This is described by

$$F(r) = \frac{K}{r^2}$$

where K is a constant. Using the Chain Rule

$$\frac{dF}{dt} = \frac{dF}{dr} \cdot \frac{dr}{dt}$$

and calculating dF/dr and dr/dt, we can determine dF/dt, the quantity we wish to find.

EXAMPLE

Satellite Problem

A satellite weighs 10 pounds on the surface of the earth. A rocket carrying the satellite is leaving the earth at the rate of 2000 mi/hr. Assume that the radius of the earth is approximately 3960 miles and that the force describing the rocket's weight varies inversely as the square of the distance from the center of the earth. Determine the rate of change of the weight of the satellite with respect to time, when the satellite is 40 miles above the surface of the earth.

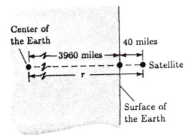

FIGURE 3.3

SOLUTION

See Figure 3.3. Let r represent the distance from the satellite to the center of the earth and let F represent the force between them. F and r are related by the equation

$$F(r) = \frac{K}{r^2}.$$

We first solve for the constant K by setting $r = 3960$ and $F = 10$.

$$10 = \frac{K}{(3960)^2}$$

$$K = 10(3960)^2$$

Thus,

$$F = \frac{10(3960)^2}{r^2}.$$

Now we use the Chain Rule to differentiate $F(r)$.

$$\frac{dF}{dt} = \frac{dF}{dr}\frac{dr}{dt} = \frac{10(3960)^2}{r^3}(-2)\frac{dr}{dt}$$

Evaluating this equation for $r = 3960 + 40 = 4000$ and $dr/dt = 2000$ produces

$$\frac{dF}{dt} = \frac{-20(3960)^2}{(4000)^3} \cdot 2000 \approx -9.8 \text{ lb/hr.}$$

Thus, the satellite is "losing" almost 10 pounds/hour.

APPLICATION 3.3 EXERCISES

1. Rework the example with the position of the satellite given by $r = -16t^2 + 21,120,000$ feet. Find dF/dt when $t = 1$.

2. The satellite in the example weighs about 1.66 pounds on the surface of the moon. If the radius of the moon is approximately 1077 miles, and if the speed of the rocket is 2000 miles/hour, find the rate of change of the weight of the satellite when the satellite is 40 miles above the surface of the moon.

3. The rocket in the example is rising at a constant velocity. If the change in the weight of the satellite is given by

$$\frac{dF}{dt} = -10 \text{ lb/hr}$$

determine the rate at which the rocket is rising when the rocket is 100 miles above the surface of the earth.

4. Consider the satellite in the example except assume that the weight of the satellite is unknown and the change in the weight of the satellite is given by

$$\frac{dF}{dt} = -20 \text{ lb/hr}$$

when the satellite is 40 miles above the surface of the earth, determine the weight of the satellite on the surface of the earth.

APPLICATION 3.4: Related Rates

> See Section **3.7**
> **Related Rates**
> **Calculus, 4th Edition,**
> **Larson/Hostetler/Edwards**

EXAMPLE

Double Shadow Problem

A man walking at the rate of 5 feet/second approaches a spotlight mounted on a wall 30 feet above the ground. The man's child begins to follow 10 feet behind him walking at the same speed. The man is 6 feet tall and the child is 3 feet tall. (See Figure 3.4.)

(a) Consider the shadow behind the child. When the man is close to the wall, the shadow is caused by the child. When the man is far from the wall, the shadow is caused by the man. Determine the distance from the man to the wall at which the man's shadow overlaps the child's. Determine the distance from the man to the wall at which the man's shadow does not overlap the child's.

(b) Determine how fast the tip of the shadow behind the child is moving as a function of the distance of the man from the base of the wall.

(c) Is the change in the position of the tip of the shadow continuous?

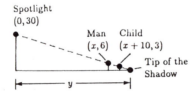

Spotlight
$(0, 30)$

Man $(x, 6)$ Child $(x + 10, 3)$

Tip of the Shadow

y

FIGURE 3.4

SOLUTION

(a) Let x represent the distance from the man to the wall. The critical distance occurs when the points $(0, 30)$, $(x, 6)$, and $(x + 10, 3)$ lie on a line. By equating the slopes, we obtain

$$\frac{30 - 6}{0 - x} = -\frac{3}{10}$$

$$3x = 240$$

$$x = 80 \text{ feet.}$$

Therefore, if $0 \leq x \leq 80$, the tip of the shadow is caused by the child. If $80 \leq x$, the shadow tip is caused by the man.

(b) Let y represent the distance from the wall to the tip of the shadow. We know that $dx/dt = -5$ and want to find dy/dt. To do this we'll consider the case in which $0 \leq x < 80$ and then the case in which $x > 80$. Let $0 \leq x < 80$. Using similar triangles, as shown in Figure 3.5, we obtain

$$\frac{y}{30} = \frac{y - x - 10}{3}.$$

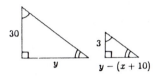

FIGURE 3.5

Thus,

$$y = 10y - 10x - 100$$
$$9y = 10x + 100.$$

Differentiation gives

$$9\frac{dy}{dt} = 10\frac{dx}{dt}$$
$$\frac{dy}{dt} = \frac{10}{9}\frac{dx}{dt} = \frac{10}{9}(-5) = \frac{-50}{9} \text{ ft/sec.}$$

The tip of the shadow behind the child is decreasing at $-50/9$ feet/second if the man is less than 80 feet from the wall. Let $80 < x$. Using similar triangles, as shown in Figure 3.6, we obtain

$$\frac{y}{30} = \frac{y - x}{6}$$
$$y = 5y - 5x$$
$$5x = 4y.$$

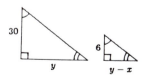

FIGURE 3.6

Differentiating gives

$$5\frac{dx}{dt} = 4\frac{dy}{dt}$$

so

$$\frac{dy}{dt} = \frac{5}{4}\frac{dx}{dt} = \frac{5}{4}(-5) = \frac{-25}{4} \text{ ft/sec}.$$

Hence,

$$\frac{dy}{dt} = \begin{cases} -\dfrac{50}{9} \text{ ft/sec}, & 0 \le x < 80 \\ -\dfrac{25}{4} \text{ ft/sec}, & 80 < x. \end{cases}$$

This means the tip of the shadow moves at two different rates depending on the value of x.

(c) Since there are two different values for dy/dt, the function is not continuous.

APPLICATION 3.4 EXERCISES

1. Consider the given example but find the change in the length of the shadow behind the child as a function of the distance of the man from the wall.
2. Consider the given example except the spotlight is placed on the ground 120 feet from the base of the wall. Find the change in the height of the shadow on the wall as a function of the distance of the man from the wall when that distance is between 0 and 100 feet.
3. Solve the given example if the height of the child is 4 feet.
4. Solve Exercise 1 assuming the height of the child is 4 feet.

APPLICATION 3.5: Melting Snow or Ice

> See Section 3.7
> **Related Rates**
> **Calculus, 4th Edition,**
> **Larson/Hostetler/Edwards**

The earth's temperature is rising, causing even the most skeptical observer concern about the effects of global warming if the process continues unchecked. Most observers believe that through cooperative reduction in the amount of carbon dioxide released by human activities; by protection and replanting of land, especially forests; by conservation of fossil fuels; by recycling; and through the application of new technologies to assist these efforts, the "worst-case scenario" can be averted.

Yet, the possible melting of either polar ice cap must be considered. We know that the melting of Antarctic ice is influenced not only by temperature, but also by such factors as existing seasonal variation in the amount of ice, wind activity, water channels, the crystalline structure of ice formed under different conditions, and the closeness of ice to land mass.*

Our problem here is a much simplified melting situation. A spherical snowball exists, formed of pure H_2O, and is melting. We are interested in investigating the volume of the snowball, as a function of time. Suppose it is known that the rate of melting is proportional to the surface area of the snowball, and that the volume can be measured at two different times. We then can determine when the snowball will be completely melted.

EXAMPLE

Snowball

A spherical snowball melts at a rate proportional to its surface area. Its volume decreases from 100 cubic centimeters at noon to 80 cubic centimeters at 1 P. M.

(a) Find the volume as a function of time.
(b) When will the snowball be completely melted?

SOLUTION

Let V represent the volume of the snowball, let S represent the surface area of the snowball, and let r represent the radius of the snowball.

(a) We have $V = \frac{4}{3}\pi r^3$. Then

$$\frac{dV}{dt} = 4\pi r^2 \frac{dr}{dt}.$$

But we also know the snowball melts at a rate proportional to its surface area. Thus,

$$\frac{dV}{dt} = KS$$

and since $S = 4\pi r^2$, we have

$$\frac{dV}{dt} = K(4\pi r^2).$$

Therefore,

$$4\pi r^2 \frac{dr}{dt} = K(4\pi r^2).$$

---

*From *Calypso Log*; 1989 (Cousteau Society).

So $dr/dt = K$ is a constant which implies that r is a linear function of t and we can write

$$r = Kt + b.$$

By substitution, we determine K and b. We let $t = 0$ represent the time at noon. Then,

$$100 = \frac{4}{3}\pi r^3 \quad \Rightarrow \quad r = \left(\frac{75}{\pi}\right)^{1/3}$$

which implies that

$$b = \left(\frac{75}{\pi}\right)^{1/3}$$

We let $t = 1$ represent the time at 1 P. M. Then,

$$80 = \frac{4}{3}\pi r^3 \quad \Rightarrow \quad r = \left(\frac{60}{\pi}\right)^{1/3}$$

which implies that

$$\left(\frac{60}{\pi}\right)^{1/3} = K(1) + \left(\frac{75}{\pi}\right)^{1/3}$$

$$K = \left(\frac{60}{\pi}\right)^{1/3} - \left(\frac{75}{\pi}\right)^{1/3}.$$

Finally, it follows that

$$r = \left[\left(\frac{60}{\pi}\right)^{1/3} - \left(\frac{75}{\pi}\right)^{1/3}\right]t + \left(\frac{75}{\pi}\right)^{1/3}$$

and so

$$V = \frac{4}{3}\pi\left(\left[\left(\frac{60}{\pi}\right)^{1/3} - \left(\frac{75}{\pi}\right)^{1/3}\right]t + \left(\frac{75}{\pi}\right)^{1/3}\right)^3.$$

(b) To determine when the snowball will be entirely melted, we can use the fact that, when melting is complete, the snowball's volume, or, equivalently its radius, will be zero.

$$\left[\left(\frac{60}{\pi}\right)^{1/3} - \left(\frac{75}{\pi}\right)^{1/3}\right]t + \left(\frac{75}{\pi}\right)^{1/3} = 0$$

$$t = \frac{(75/\pi)^{1/3}}{(75/\pi)^{1/3} - (60/\pi)^{1/3}} \approx 14 \text{ hours.}$$

Thus, the snowball will be completely melted in about 14 hours.

APPLICATION 3.5 EXERCISES

1. Solve the problem as given in the Example except assume that the volume is 125 cubic centimeters at noon and 64 cubic centimeters at 2 P. M.

2. An ice cube in the shape of a cube melts at a rate proportional to its surface area. If the volume was 125 cubic centimeters at noon and 64 cubic centimeters three hours later, determine the volume of the ice cube as a function of time. When will the ice cube be completely melted?

3. An ice cube in the shape of a cube melts at a rate proportional to its surface area. If the volume was 125 cubic centimeters at noon and if the ice cube was completely melted in six hours, determine the volume of the ice cube at 1 P. M.

4. An icemaker produces ice in the shape of half disks. (See figure.) If the radius of a piece of ice was 3 centimeters at noon and 2 centimeters thirty minutes later, determine the volume of the ice piece as a function of time. Assume that the height of the ice piece is equal to $\frac{2}{3}$ the radius.

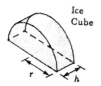

FIGURE FOR EXERCISE 4

4

APPLICATIONS OF DIFFERENTIATION

APPLICATION 4.1: Maximum Illumination

> **See Section 4.1**
> **Extrema on an Interval**
> **Calculus, 4th Edition,**
> **Larson/Hostetler/Edwards**

Wherever you live, you experience a difference in the amount and intensity of sunlight as the seasons change. The intensity of the sun's rays depends on the angle at which the rays meet the surface of the earth. It is the variation both in the distance and the angle from the sun to the locality that provides our familiar seasonal variations.

Indoors we have similar relationships between a light source and the lighted object. The amount of illumination on a surface is *proportional to* the intensity of the light source, is *inversely proportional to* the square of the distance from the light source, and is *proportional to* $\sin \theta$, where θ is the angle at which the light strikes the surface.

EXAMPLE

A Lighted Room

A rectangular family room measures 10 feet by 24 feet, with an 8-foot ceiling. A light is to be suspended from the center of the ceiling. Determine the height at which the light should be placed to allow the corners at the floor to receive as much light as possible.

SOLUTION

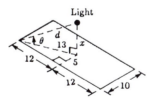

FIGURE 4.1

Let x represent the height of the light above the floor, let θ represent the angle at which the light strikes the surface, and let I represent the illumination at the corners. (See Figure 4.1.) Since $\sin \theta = x/d$, we obtain

$$I = \frac{k \sin \theta}{d^2} = \frac{kx}{d^3}.$$

Using $d^2 = x^2 + 13^2$, we obtain

$$I = \frac{kx}{(x^2 + 169)^{3/2}}, \quad 0 \le x \le 8.$$

We wish to maximize I on the interval $[0, 8]$. We begin by setting $dI/dx = 0$ as follows

$$\frac{dI}{dx} = k\left[\frac{(x^2 + 169)^{3/2} - x(3/2)(x^2 + 169)^{1/2}(2x)}{(x^2 + 169)^3}\right]$$

$$= k(x^2 + 169)^{1/2}\frac{(x^2 + 169) - 3x^2}{(x^2 + 169)^3} = 0.$$

Thus,

$$2x^2 = 169$$

$$x^2 = \frac{169}{2}$$

$$x = \frac{13}{\sqrt{2}} \approx 9.2.$$

which is not in the interval. We now check the endpoints, beginning with $I(0) = 0$. Clearly, this is not the maximum, so

$$I(8) = \frac{k8}{(64 + 169)^{3/2}} \approx 0.0022k$$

must be the maximum. Thus, the light should be placed flush with the ceiling. (Note that the particular dimensions of the room are important.)

APPLICATION 4.1 EXERCISES

1. Solve the given problem when the ceiling is 10 feet high.

2. Solve the given problem when the family room measures 18 feet by 24 feet, with a 12-foot ceiling.

3. Solve the given problem using θ as the independent variable. At what angle θ will the illumination in the corners be maximum?

4. A square family room of unknown length has a ceiling that is higher than 8 feet. The illumination at the corners of the room is a maximum when the light is 8 feet above the floor. Determine the dimensions of the room.

APPLICATION 4.2: Projectiles and Range

> See Section **4.3**
> **Increasing and Decreasing Functions**
> **   and the First Derivative Test**
> **Calculus, 4th Edition,**
> **Larson/Hostetler/Edwards**

Suppose you and a friend are competing to see who can throw a stone the farthest. You are standing on level ground. In this classical projectile problem, the maximum range of the stone (distance the stone goes before hitting the ground) can be described in terms of its initial velocity.

Here we consider the more complex case in which the game is played on a hill. Figure 4.2 illustrates the situation: θ is the angle of inclination of the hill, the projectile has an initial velocity v_0, and we assume no air resistance. Two angles, θ and ϕ, affect the distance x the projectile can reach.

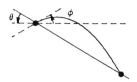

FIGURE 4.2

EXAMPLE
━━━━━━━

Downhill Toss

For a downhill stone toss, what is the angle ϕ that permits the stone to travel the farthest?

SOLUTION

In order to solve the problem, we must express the impact coordinates in terms of the angle θ. In Figure 4.2 we see that we can decompose the x and y distances by using the angles θ and ϕ. The results are

$$x = v_0(\cos \phi)t$$

and

$$y = v_0(\sin \phi)t - \frac{1}{2}gt^2 = -x\tan \theta.$$

We solve for t in the x-equation.

$$t = \frac{x}{v_0 \cos \phi}$$

Substituting this value into the y-equation, we obtain

$$-x \tan\theta = \frac{xv_0 \sin\phi}{v_0 \cos\phi} - \frac{gx^2}{2(v_0 \cos\phi)^2}$$

$$-x \tan\theta - x \tan\phi = -\frac{gx^2}{2(v_0 \cos\phi)^2}.$$

The value $x = 0$ represents a minimum distance and is not the solution we seek. Thus, we can divide both sides of the equation by x and solve the remaining equation for x. This leads to

$$x = \frac{2v_0^2 \cos^2\phi (\tan\phi + \tan\theta)}{g}$$

$$= k(\cos\phi \sin\phi + \cos^2\phi \tan\theta)$$

where $k = 2v_0^2/g > 0$. At this point we have an equation expressing x in terms of the two angles θ and ϕ. We seek those values that will maximize x. We differentiate with respect to ϕ. The derivative is

$$\frac{dx}{d\phi} = k(\cos 2\phi - \sin 2\phi \tan\theta)$$

which yields a critical number at

$$\cot 2\phi = \tan\theta.$$

Using a cofunction identity, we obtain

$$\tan\left(\frac{\pi}{2} - 2\phi\right) = \tan\theta$$

$$\frac{\pi}{2} - 2\phi = \theta$$

$$2\phi = \frac{\pi}{2} - \theta$$

$$\phi = \frac{\pi}{4} - \frac{\theta}{2}.$$

We conclude that $\phi = (\pi/4) - (\theta/2)$ for maximum range. Notice that the result reduces to the maximum range result for a horizontal situation when θ approaches zero.

APPLICATION 4.2 EXERCISES

1. Solve a similar problem in which the stone is thrown uphill on a hill with angle of inclination θ.

2. In the given example, assume that the angle of inclination of the hill is $\theta = 20°$. How far downhill can an object whose initial velocity is 64 feet/second be thrown?

3. You are throwing stones off a 100-foot cliff overlooking a lake. At what angle should you throw your stone to achieve the maximum range? Again, assume no air resistance. Also, assume that the initial velocity of the object is 64 feet/second.

4. You and your friends are having a snowball fight. The teams are located on opposite sides of a flat-roofed garage twelve feet tall and twenty feet wide. Suppose the other team throws snowballs at a speed of 64 feet/second. Is there a "safe zone" near the garage in which you can be certain that the other team will not be able to hit you with a snowball? Is so, describe this zone. (Assume the snowballs are thrown from a height of 6 feet.)

APPLICATION 4.3: Optical Paths

See Section **4.3**
**Increasing and Decreasing Functions
   and the First Derivative Test**
Calculus, 4th Edition,
Larson/Hostetler/Edwards

EXAMPLE

Optical Path

A bright young scientist is on board a small research submarine. The viewing port is a hemispherical dome, curving in toward the inside of the submarine. The scientist recognizes that the water-air system makes an interesting lens system. She considers the following situation. A fish swimming along the axis of the lens at a distance d from the water-air interface looks into the viewing port and sees a worm. Will the optical path between the fish and a point x on the lens inside the submarine be a critical point, as suggested by Fermat's Principle? (Ignore the window thickness.) See Figure 4.3.

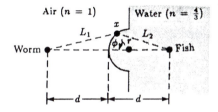

FIGURE 4.3

SOLUTION

We will see that the optical path (the path which the light ray will travel) is either a minimum, an inflection point, or a maximum. To determine the path, we begin with a ray close to the expected straight-line path, where the angle ϕ is small. Using the Law of Cosines, the lengths L_1 and L_2 are given by

$$L_1 = \sqrt{(d+r)^2 + r^2 - 2r(d+r)\cos\phi}$$

and

$$L_2 = \sqrt{(d-r)^2 + r^2 + 2r(d-r)\cos\phi}.$$

The time it takes for a ray to travel from the worm A to the fish eye B is then

$$t = \frac{L_1 + nL_2}{c}$$

where n is the index of refraction of the water, and c is the speed of light. This gives us the expression

$$t = \frac{1}{c}\left(\sqrt{(d+r)^2 + r^2 - 2r(d+r)\cos\phi} + n\sqrt{(d-r)^2 + r^2 + 2r(d-r)\cos\phi}\right).$$

To find whether we have a maximum, a minimum, or an "inflection" solution, we use the first derivative

$$\frac{dt}{d\phi} = \frac{1}{c}\left(\frac{-r(d+r)\sin\phi}{\sqrt{(d+r)^2 + r^2 - 2r(d+r)\cos\phi}} + \frac{nr(d-r)\sin\phi}{\sqrt{(d-r)^2 + r^2 + 2r(d-r)\cos\phi}} \right).$$

For small ϕ, the denominators are nearly equal. Hence, for values of ϕ near zero, we can rewrite $dt/d\phi$ as

$$\frac{dt}{d\phi} = \frac{1}{c}\left[\frac{-r(d+r)\sin\phi + nr(d-r)\sin\phi}{d} \right]$$

$$= \frac{r}{cd}[d(n-1) - r(n+1)]\sin\phi$$

$$= \frac{r}{cd}\left[d - \frac{r(n+1)}{n-1} \right](n-1)\sin\phi.$$

We see that for $\phi = 0$, we do indeed have a critical number.

APPLICATION 4.3 EXERCISES

1. Show that for $d > \dfrac{r(n+1)}{n-1}$, the length of the optical path is greater than the "inflection" solution.

2. Show that for $d = \dfrac{r(n+1)}{n-1}$, the optical path corresponds to an inflection point. This means that small changes in ϕ will not change the sign of the first derivative. [Note: Inflection points are also called *stationary* points.]

3. Show that for $d < \dfrac{r(n+1)}{n-1}$, the length of the optical path is less than the "inflection" solution.

4. Show that $L_1 = L_2$ when $\phi = 0$.

APPLICATION 4.4: Focal Length

> See Section 4.5
> Limits at Infinity
> Calculus, 4th Edition,
> Larson/Hostetler/Edwards

Light rays coming from infinity and passing through a converging lens will converge (or meet) at the focus on the opposite side of the lens. The distance from the lens to the focus is called the focal length. (See Figure 4.4.)

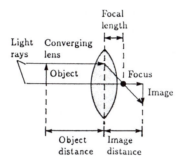

FIGURE 4.4

To determine the focal length of the system in Figure 4.4, we measure the object distance d_o (the distance from an object to the lens) and the image distance d_i (the distance from the image to the lens). Then, plotting the object distance on the vertical axis and the image distance on the horizontal axis, we sketch the graph shown in Figure 4.5.

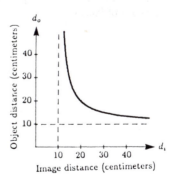

FIGURE 4.5

The curve sketched from this data is a hyperbola. We expect that the horizontal and vertical asymptotes yield the focal length because the asymptotes would correspond to having either the source or the image at infinity.

EXAMPLE

Focal Length

By examining the data in Figure 4.5 from another perspective, determine the focal length of the lens shown in Figure 4.4.

SOLUTION

Another way to examine these data is to plot $1/d_o$ versus $1/d_i$. If the original curve is a hyperbola, then the new graph should be a straight line as shown in Figure 4.6. Each intercept is the reciprocal of the focal length. Thus, the focal length of the lens in Figure 4.4 is 10 centimeters.

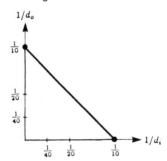

FIGURE 4.6

APPLICATION 4.4 EXERCISES

1. A diverging lens is placed next to the converging lens described in the example. The relationship between the object distance and the image distance for the lens system is shown in Figure 4.7. Find the effective focal length of this system of lenses.

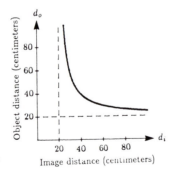

FIGURE 4.7

2. Table 4.1 represents the height and the distance a projectile travels when shot from a laboratory cannon.

TABLE 4.1

Height (cm)	10	20	30	40	50	60
Distance (cm)	28.6	40.4	49.5	57.1	63.9	70.0

(a) It is believed that the relationship between the height and the distance is

$$D = v_0 \sqrt{\frac{2H}{g}}.$$

Plot D versus $\sqrt{H}$ to verify this relationship.

(b) Verify that the slope of this line is $v_0 \sqrt{2/g}$.

(c) Use the slope of the curve to determine the value of v_0.

3. A series-wound, variable-speed, $\frac{1}{2}$-horsepower, electric drill is used in some drilling processes. The data graphed in Figure 4.8 shows the drill output torque τ and the current I that the drill uses at different torques. Determine the relationship between the torque and the current. Assume $\tau = kI^2$. Plot τ versus I^2. If the result is a straight line, use the slope to determine the value of k.

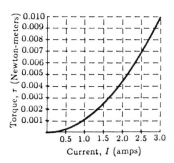

FIGURE 4.8

APPLICATION 4.5: Doppler Shift

> **See Section 4.7**
> **Optimization Problems**
> **Calculus, 4th Edition,**
> **Larson/Hostetler/Edwards**

The Doppler shift is the change in frequency of sound due to the motion of the observer or the sound source. In the traditional textbook presentation, this motion is along a line joining the source and observer. However, with no minimum separation defined between them, the observer and the sound source can collide with each other!

A more realistic discussion of Doppler shift might include a variable called the *impact parameter d*. This is defined as the closest distance ever achieved between the source and the observer.

Here we will develop two expressions for the Doppler shift: in (1) the sound source is fixed and only the observer moves, while in (2) both source and observer move. For our analysis we take v_o and v_s as the speeds of the observer and source, respectively, and we let c be the speed of sound (1100 feet/second at room temperature).

(1) See Figure 4.9. Here the observer moves in the x direction, forming an angle θ with the source. The Doppler shift is given by

$$f = \frac{f_0(c + v_o \cos\theta)}{c}.$$

Notice the symmetry. On each side of the closest distance d, there is the same frequency shift.

FIGURE 4.9   FIGURE 4.10

(2) See Figure 4.10. For the second case, both the source and the observer move. Here, ϕ is the angle the moving sound source makes with the direction toward the observer. The Doppler shift is given by

$$f = \frac{f_0(c + v_o \cos\theta)}{c - v_s \cos\phi}.$$

EXAMPLE

Burglar Alarm

Driving along at 30 miles/hour (44 feet/second), you pass a sounding burglar alarm that is emitting a pure tone of 400 Hertz. (See Figure 4.11.)

(a) How far away from the alarm are you if the apparent frequency shifts from 408 Hertz to 392 Hertz in one second?

(b) What is the rate at which the frequency is changing when you are closest to the source?

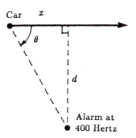

FIGURE 4.11

SOLUTION

This is Case (1) in which the observer only is moving.

(a) Since the moving observer case is symmetric, we know that there is a frequency shift of 8 Hertz on each side of the closest point. In $\frac{1}{2}$ second then, we would travel 22 feet. We have the Doppler shift

$$408 = 400 \left(\frac{1100 + 44\cos\theta}{1100} \right)$$

which gives us a value of

$$\cos\theta = \frac{(408)(1100) - (400)(1100)}{44(400)}$$

$$\cos\theta = \frac{1}{2}$$

$$\theta = \frac{\pi}{3}.$$

Thus, the impact parameter (closest point) is $d = 22\sqrt{3}$ feet.

(b) The rate at which the frequency is changing at the closest point is obtained from the derivative of

$$f = \frac{f_0(c + v_o \cos \theta)}{c}$$

which is

$$\frac{df}{dt} = -\frac{f_0 v_o}{c} \sin \theta \, \frac{d\theta}{dt}.$$

From Figure 4.11 we see that $x/d = \cot \theta$. This gives us

$$\frac{d\theta}{dt} = -\frac{\sin^2 \theta}{d} \frac{dx}{dt}.$$

Recall that $dx/dt = -v_o = -44$ ft/sec and $\theta = \pi/2$. Thus,

$$\frac{df}{dt} = \frac{f_0 v_o \sin^3 \theta}{cd} \frac{dx}{dt} = \frac{f_0 v_o^2 \sin^3 \theta}{cd}.$$

Therefore, we get $\dfrac{df}{dt} = -18.48$ Hz/sec.

APPLICATION 4.5 EXERCISES

1. In the example given, verify that the expression

$$f = \frac{f_0(c + v_o \cos \theta)}{c}$$

reduces to the typical physics textbook value

$$f = f_0 \left(1 \pm \frac{v_o}{c}\right)$$

when the impact parameter approaches zero.

2. In the example given, where is the frequency changing most rapidly? What is the maximum change that could occur?

3. Show that the expression

$$f = f_0 \left(\frac{c + v_o \cos \theta}{c - v_s \cos \phi}\right)$$

reduces to the classic physics text expression

$$f = \frac{f_0(c \pm v_o)}{c \mp v_s}$$

when the impact parameter approaches zero.

4. A freight train is traveling toward you at 88 feet/second, with an impact parameter of 100 feet. The whistle on the train is blown with a frequency of 440 Hertz. Find the rate at which the frequency is changing when the engine (whistle source) is directly in front of you (and as close to you as possible). Recall that the impact parameter is the closest distance between the source and observer.

APPLICATION 4.6: Using Optics to Find the
Concentration of Sugar

> See Section **4.7**
> Optimization Problems
> Calculus, 4th Edition,
> Larson/Hostetler/Edwards

Manufacturers of jellies and wines must accurately measure the concentration of sugar in their products. Transparent liquids in prism-shaped glass containers lend themselves to optical measurement. (See Figure 4.12.)

FIGURE 4.12   **FIGURE 4.13**

The underlying physics needed is Snell's Law (see Figure 4.13)

$$n_1 \sin \theta_1 = n_2 \sin \theta_2$$

and the underlying mathematics is determining a minimum.

Let's assume the prism is triangular, with one vertex, a right angle. When we pass a light ray through the prism, the minimum deflection of the ray will accurately measure the index of refraction of the liquid within the container. Since the prism is surrounded by air, let $n_1 = 1$ and $n_2 = n$, the index of refraction of the liquid.

EXAMPLE

Ray Deflection and Percentage of Sugar

Suppose a light ray is passed through a triangular prism containing a transparent medium. (See Figure 4.14.) Here δ represents the angle of deflection of the ray as it passes through the prism. If small amounts of sugar are added to the liquid, there will be a slight change in the index of refraction. Use a prism and Snell's Law to obtain a relationship for the index of refraction, n, in terms of the angle of deflection, δ.

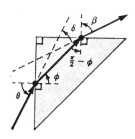

FIGURE 4.14

SOLUTION

First we will determine the minimum deflection for a ray, as shown in Figure 4.14. Once this is determined, we will use the differential to determine the change in index of refraction when the change is small. From the diagram and Snell's Law, we have

$$\sin \theta = n \sin \phi$$

$$n \sin \left(\frac{\pi}{2} - \phi \right) = \sin \beta \qquad \Longrightarrow \qquad n \cos \phi = \sin \beta.$$

Squaring these two equations yields

$$\sin^2 \theta = n^2 \sin^2 \phi$$
$$\sin^2 \beta = n^2 \cos^2 \phi$$

and by adding the results, we obtain

$$\sin^2 \theta + \sin^2 \beta = n^2.$$

We also have

$$\delta = \theta - \left(\frac{\pi}{2} - \beta \right) = \theta + \beta - \frac{\pi}{2}.$$

Differentiating with respect to θ, we obtain

$$\delta' = 1 + \beta' \quad \text{and} \quad (2 \sin \theta \cos \theta) + (2 \sin \beta \cos \beta) \beta' = 0.$$

Thus,

$$\beta' = - \frac{\sin 2\theta}{\sin 2\beta}$$

$$\delta' = 1 - \frac{\sin 2\theta}{\sin 2\beta} = \frac{\sin 2\beta - \sin 2\theta}{\sin 2\beta}.$$

Critical points occur when $\beta = \theta$, which means that

$$\phi = \frac{\pi}{2} - \phi = \frac{\pi}{4}.$$

This result is independent of the liquid within the prism since n does not appear in this expression. We currently have $n > 1$, $\phi = \pi/4$, and $\beta = \theta$, which leads to

$$2 \sin^2 \theta = n^2$$

$$\sqrt{2} \sin \theta = n.$$

However, we know that

$$\delta = 2\theta - \frac{\pi}{2}$$

$$\frac{\delta + (\pi/2)}{2} = \theta$$

$$n = \sqrt{2} \sin \left[\frac{\delta + (\pi/2)}{2} \right].$$

This gives the index of refraction in terms of the angle of deflection δ. By measuring the minimum deflection of a ray passing through a prism, we can make an accurate determination of the index of refraction.

APPLICATION 4.6 EXERCISES

1. A small change in the index of refraction (caused perhaps by adding sugar to water) is given by the differential. Use the result in the example to show that

$$dn = \frac{1}{2}\left(\cos\frac{\delta}{2} - \sin\frac{\delta}{2}\right) d\delta.$$

2. Find δ for water $(n = 1.333)$.

3. Complete Table 4.2 for sugar dissolved in water.

   TABLE 4.2

Sugar (%)	10	20	40
δ (deg)	54.8	59.4	73.7
n	a	b	c

4. Find δ associated with a 30% sugar solution. (See Exercise 3 and Table 4.2.)

APPLICATION 4.7: Relativistic Mass

> See Section 4.9
> Differentials
> Calculus, 4th Edition,
> Larson/Hostetler/Edwards

According to the Special Theory of Relativity, the relativistic mass of an object increases as its velocity increases. We shall assume the formula

$$m(v) = \frac{m_0}{\sqrt{1 - (v^2/c^2)}} = m_0 \left(1 - \frac{v^2}{c^2}\right)^{-1/2}$$

where m_0 represents the rest mass of the object, c represents the speed of light, and $m(v)$ represents the mass of that object relative to a stationary observer. We wish to measure the increase in mass.

The velocities that we generally encounter are so small that this increase in mass is essentially not measurable. However, particle accelerators such as the Fermilab in Batavia, Illinois are capable of accelerating particles to near the speed of light. At these velocities the increase in mass is significant and, as a result, an enormous amount of energy is required for acceleration.

EXAMPLE

Particle Acceleration

Assume a particle's velocity increases to nearly the speed of light and the relativistic mass is described by $m(v)$. Use the differential to estimate the change in the relativistic mass of the object as the velocity of that object increases from $0.9c$ to $0.91c$.

SOLUTION

Since

$$m(v) = m_0 \left(1 - \frac{v^2}{c^2}\right)^{-1/2}$$

then

$$\Delta m \approx dm = \frac{dm}{dv} dv$$

$$= m_0 \left(-\frac{1}{2}\right) \left(1 - \frac{v^2}{c^2}\right)^{-3/2} \left(-\frac{2v}{c^2}\right) dv.$$

We evaluate this for $v = 0.9c$ and $dv = 0.91c - 0.9c = 0.01c$.

$$dm = m_0 \left(-\frac{1}{2}\right)(1 - 0.81)^{-3/2}(-2)(0.9)(0.01)$$

$$\approx 0.11 m_0$$

Thus, the change in relativistic mass corresponding to a change in velocity from $0.9c$ to $0.91c$ is approximately one-ninth the rest mass of the particle!

APPLICATION 4.7 EXERCISES

1. Solve the given problem for the case when the velocity increases from $0.1c$ to $0.11c$.

2. An object is moving at velocity $0.8c$. Determine the approximate velocity necessary to make the relativistic mass increase by 20%.

3. Assuming $E = mc^2$, use the differential to approximate the change in energy as the velocity of an electron increases from $0.8c$ to $0.81c$. Assume that the rest mass of an electron $m_0 = 9.1095 \times 10^{-31}$ kilograms and the speed of light $c = 2.9979 \times 10^8$ meters/second.

4. Solve Exercise 3 but assume that the accelerated particle is a proton of rest mass $m_0 = 1.6726 \times 10^{-27}$ kilograms.

5. Rework the given example to estimate the change in the relativistic mass of the object as the velocity of the object increases form $0.9c$ to $0.95c$. Compute both Δm and dm. Why does the differential yield such a poor approximation here?

APPLICATION 4.8: Resonance Model

> See Section 4.9
> Differentials
> Calculus, 4th Edition,
> Larson/Hostetler/Edwards

Suppose we have a simple tube—long, thin, and closed-ended. If we blow across the open end of the tube we get a tone, a resonance with a fundamental frequency, and an entire overtone series. This resonance tube thus supports many frequencies, each of which relate in a simple way to the fundamental lowest frequency. This type of tube provides a simple model upon which more sophisticated models can be based.

EXAMPLE

Modeling Soda Bottle Resonance

Blowing across the mouth of a glass soda bottle also gives a tone, but the soda bottle, unlike the resonance tube, can only support one frequency. Consider some assumptions for the soda bottle situation. (1) Assume the air in the neck of the bottle remains intact as a moving mass, and the restoring force is due to the mass compressing the air adiabatically (without flow of heat) within the bottle. (2) Also assume the bottle is glass, full of air, and in a room at atmospheric pressure. Using Hooke's Law $(F = -kx)$, develop a model for the soda bottle as a resonator.

SOLUTION

Let F be the force on the gas in the body of the bottle. Then the force on the mass trapped within the *neck* of the bottle is

$$dF = A \, dP$$

where A is a cross-sectional area of the neck and dP is the change in pressure from the equilibrium pressure (atmospheric pressure). Since the gas is expanded adiabatically, it obeys the law

$$PV^\gamma = \text{constant}$$

where γ is a constant that depends on the structure of gas. (For air, $\gamma \approx 1.4$.) Using differentials, we get

$$dP \, V^\gamma + \gamma V^{\gamma-1} P \, dV = 0$$

$$dP = -\gamma V^{\gamma-1} \frac{P \, dV}{V^\gamma} = -\frac{\gamma P \, dV}{V}.$$

However, for the neck of the bottle $dV = A \, dx$, so

$$dF = A \, dP = -\frac{\gamma P A^2 \, dx}{V} \quad \Longrightarrow \quad F = -\frac{\gamma P A^2}{V} x = -kx.$$

The mass of the gas in the neck is $m = \rho A L$, where L is the length of the neck, and ρ is the density of the gas within the neck. This provides a way of obtaining ω, the angular frequency for the resonance

$$\omega = \sqrt{\frac{k}{m}} = \sqrt{\frac{\gamma P A^2}{\rho A L V}} = \sqrt{\frac{\gamma P A}{\rho L V}}$$

and f, the frequency for the resonance

$$f = \frac{\omega}{2\pi}.$$

This model describes the bottle resonant frequency f well until the bottle volume V becomes smaller and close to the volume within the neck of the bottle. As the volume of the gas in the body approaches the volume of the neck or becomes even smaller than that volume, the physical system looks more and more like a uniform resonance tube with one end closed.

APPLICATION 4.8 EXERCISES

1. Suppose the process is isothermal instead of adiabatic. Then we can assume there is no change in temperature. We also assume the gas within the body of the bottle is trapped by a non-interactive membrane between the neck and the body. (a) Calculate the bottle's resonant frequency, assuming that $PV = nRT = $ constant. (b) Compare the results of this exercise to those of the example. Do the two models predict similar behavior?

2. In the example, the only expression that is temperature-dependent is the density of the gas in the bottle neck. Recall that our example model was for a glass bottle full of air in a room at atmospheric pressure. (a) Now develop a model for the density of the mass within the neck as a function of temperature. (b) Calculate the change in frequency with a 10° Celsius change in temperature. *Hints:* Recall that temperatures for gas problems are always given in degrees Kelvin. P and V can be assumed to be constant. (c) Does the frequency increase or decrease with increases in temperature? Is this the expected result?

3. The displacement of air in a cylindrical column will obey the wave equation. The speed of propagation is then $\sqrt{B/\rho}$, where B is the bulk modulus, ρ is the density, and c is the speed of propagation. For air at room temperature, $c = 343$ meters/second. Displacement waves will have antinodes at the ends of the open tubes when the injected sound source is at the resonance frequency of the tube. (The actual antinode is slightly outside the tube. We shall ignore this correction.) Such a wave with nodes and antinodes at fixed points in space is called a standing wave.

   (a) Show that a standing wave results from adding two periodic sinusoidal waves, each traveling in the opposite direction. Let the waves be given by $y_1 = A\sin(kx + ct + \phi)$ and $y_2 = A\sin(kx - ct + \phi)$.

   (b) For the open-ended resonance tube, show that $\phi = \pi/2$ if one end is labeled 0 and the other end is labeled L.

   (c) Find the allowable k values for resonance. Use the relationship $\lambda f = c$, where λ is the wavelength of the wave, to find the allowed resonance frequencies (overtone series) if the resonance tube is 1 meter long.

   (d) Suppose one end of the resonance tube is closed. What will be the appropriate boundary condition for this displacement wave?

   (e) At what length will a closed resonance tube in part (d) have the same fundamental (lowest) resonance frequency as the open resonance tube in part (c)? What is the resulting overtone series?

5

INTEGRATION

APPLICATION 5.1: Numerical Integration

> See Section 5.6
> **Numerical Integration**
> **Calculus, 4th Edition,**
> **Larson/Hostetler/Edwards**

We like to think that the stretching of an object such as a spring can be described completely using Hooke's Law, $F = -kx$. This is an overly simplistic view. Many physical situations that involve stretching a spring only produce the predicted linear displacement for small amounts of applied force. In such situations we experimentally measure the actual displacements produced by the entire range of forces and analyze the resulting table of values using numerical techniques. For example, we can use *numerical integration* to answer the question of how much energy is stored in a stretched object that is behaving nonlinearly.

EXAMPLE

A Stretched Rubber Band

When we stretch a rubber band we might think the displacement and force obey Hooke's Law $(F = -kx)$. However, this isn't true. Table 5.1 shows data obtained experimentally for the amount a rubber band is stretched versus the force applied.

TABLE 5.1

Displacement (m)	0	0.01	0.02	0.03	0.04	0.05
Force (newtons)	0	0.05	0.10	0.15	0.19	0.24

Displacement (m)	0.06	0.07	0.08	0.09	0.10
Force (newtons)	0.28	0.32	0.35	0.38	0.40

For small displacements, Hooke's Law is obeyed. This can be seen by considering the first three data points. Each additional 0.05 newtons stretches the rubber band an additional 0.01 meters. This is the linear relationship expressed in Hooke's Law. If the rubber band were to continue to stretch as predicted by Hooke's Law, it would stretch to 0.08 meters with 0.40 newtons instead of 0.10 meters.

The question at hand is: "How much energy is stored in the rubber band as a function of displacement?"

SOLUTION

We will obtain the solution to this problem through integration. The stored energy is given by the area under the curve when force is plotted against displacement. The work done in stretching the rubber band is essentially the same as the energy stored in the rubber band (assuming the stretching does not tear or heat up the rubber band).

We can obtain reasonable results for the area under the curve by the Trapezoidal Rule. The Trapezoidal Rule uses the average height of each interval times the width. Our widths are each 0.01 meters, and the areas for each successive trapezoid are as given.

0.00025	0.00075	0.00125	0.0017	0.00215
0.0026	0.0030	0.00335	0.00365	0.0039

Table 5.2 shows the displacement and energy, calculated by numerical integration using the Trapezoidal Rule.

TABLE 5.2

Displacement (m)	0	0.01	0.02	0.03	0.04	0.05
Energy (joules)	0	0.00025	0.0010	0.00225	0.00395	0.0061

Displacement (m)	0.06	0.07	0.08	0.09	0.10
Energy (joules)	0.0087	0.0117	0.0151	0.0187	0.0226

Experiments actually show that more than 90% of the stored energy in a stretched rubber band is lost due to air resistance if the rubber is shot into the air!

APPLICATION 5.1 EXERCISES

1. Use the data from Table 5.2 to plot the stored energy versus the displacement for our sample rubber band.
2. Use Simpson's Rule to obtain the areas in each 2-centimeter interval and then collectively add them as appropriate to get the total work as a function of displacement.
3. Use the data in Exercise 2 to (a) plot the stored energy versus displacement and (b) sketch the curve.
4. A nonlinear spring obeys the force law

$$F = 100(\sqrt{1 + 0.1x} - 1).$$

For small displacements, this spring appears to obey Hooke's Law. Find the spring constant for small displacements.

5. For the nonlinear spring of Exercise 4, find the energy stored in the spring when it is stretched up to 1 meter in length. Use any numerical technique and obtain values for stored energy for each 0.1-meter interval of displacement.

APPLICATION 5.2: Fourier Series and Temperature

> See Section 5.6
> **Numerical Integration**
> **Calculus, 4th Edition,**
> **Larson/Hostetler/Edwards**

A common problem in experimentation involves taking data and finding the function that "best fits" this data. For data that appears to be linear or exponential one would try to find the appropriate linear or exponential function. If the data is periodic in nature, one would try to use periodic functions. Some examples of data that appear to be periodic are rainfall quantities, amount of sunspot activity, and amount of earthquake activity.

Temperature readings for a given locality also appear to be periodic with a period of 12 months. We can approximate temperature by a periodic function comprised of sines and cosines, that is, by a Fourier series. The task centers about determining the coefficients in the series.

EXAMPLE

Temperatures in Auckland, New Zealand

Table 5.3 shows the average maximum temperatures for Auckland, New Zealand. Find the function that best fits this data.

TABLE 5.3

January	79° F	May	67° F	September	65° F
February	79° F	June	63° F	October	68° F
March	77° F	July	62° F	November	73° F
April	73° F	August	63° F	December	76° F

SOLUTION

Let t represent the time in months (January is $t = 0$) and let $f(t)$ represent the temperature in degrees Fahrenheit. We assume that $f(t)$ can be approximated by a Fourier series,

$$f(t) \approx a_0 + a_1 \cos \frac{\pi t}{6} + b_1 \sin \frac{\pi t}{6}$$

and

$$a_0 = \frac{1}{12} \int_0^{12} f(t)\, dt$$

$$a_1 = \frac{1}{6} \int_0^{12} f(t) \cos \frac{\pi t}{6}\, dt$$

$$b_1 = \frac{1}{6} \int_0^{12} f(t) \sin \frac{\pi t}{6}\, dt.$$

Note that these sines and cosines have a period of 12. We use the Trapezoidal Rule with $n = 12$ to approximate these coefficients. We find the value of a_0 as follows.

$$\int_0^{12} f(t)\, dt \approx \frac{12}{24}[79 + 2(79) + 2(77) + 2(73) + 2(67) + 2(63) + 2(62)$$

$$+ 2(63) + 2(65) + 2(68) + 2(73) + 2(76) + 79]$$

$$= \frac{1}{2}(1690) = 845$$

So $a_0 = \frac{845}{12} \approx 70.4$. We find the value of a_1 as follows.

$$\int_0^{12} f(t) \cos \frac{\pi t}{6}\, dt \approx \frac{12}{24}\left[79 + 2(79) \cos \frac{\pi}{6} + 2(77) \cos \frac{\pi}{3} + 2(73) \cos \frac{\pi}{2}\right.$$

$$+ 2(67) \cos \frac{2\pi}{3} + 2(63) \cos \frac{5\pi}{6} + 2(62) \cos(\pi)$$

$$+ 2(63) \cos \frac{7\pi}{6} + 2(65) \cos \frac{4\pi}{3} + 2(68) \cos \frac{3\pi}{2}$$

$$\left. + 2(73) \cos \frac{5\pi}{3} + 2(76) \cos \frac{11\pi}{6} + 79\right]$$

$$\approx \frac{1}{2}(102.2) = 51.1$$

So $a_1 = \frac{1}{6}(51.1) \approx 8.52$. Using a similar argument, the value of b_1 is

$$\int_0^{12} f(t) \sin \frac{\pi t}{6}\, dt \approx 11.7.$$

So, $b_1 = \frac{1}{6}(11.7) \approx 1.95$. We now have the coefficients we seek for the Fourier series. Thus,

$$f(t) \approx 70.4 + 8.52 \cos \frac{\pi t}{6} + 1.95 \sin \frac{\pi t}{6}.$$

Table 5.4 and Figure 5.1 compare the actual data with the values provided by this Fourier approximation. In practice more terms would be used and more efficient techniques for approximating the coefficients would be applied. Better approximations would produce an even better fit than that we have developed here.

TABLE 5.4

t	0	1	2	3	4	5
$f(t)$	78.9	78.8	76.3	72.4	67.8	64.0
Actual Temperature	79	79	77	73	67	63

t	6	7	8	9	10	11
$f(t)$	61.9	62.0	64.5	68.5	73.0	76.8
Actual Temperature	62	63	65	68	73	76

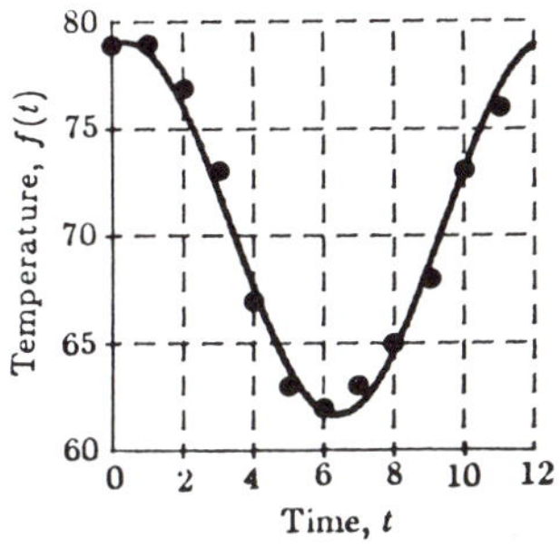

FIGURE 5.1

APPLICATION 5.2 EXERCISES

1. If $f(t) = a_0 + a_1 \cos \dfrac{\pi t}{6} + b_1 \sin \dfrac{\pi t}{6}$, show that

$$a_0 = \frac{1}{12}\int_0^{12} f(t)\,dt, \qquad a_1 = \frac{1}{6}\int_0^{12} f(t)\cos\frac{\pi t}{6}\,dt, \qquad \text{and} \qquad b_1 = \frac{1}{6}\int_0^{12} f(t)\sin\frac{\pi t}{6}\,dt.$$

2. For $f(t)$ as in the example, use Simpson's Rule to estimate a_0, a_1, and b_1.

3. If $f(t) = a_0 + a_1 \cos \dfrac{\pi t}{6} + b_1 \sin \dfrac{\pi t}{6} + a_2 \cos \dfrac{\pi t}{3} + b_2 \sin \dfrac{\pi t}{3}$, show that

$$a_2 = \frac{1}{6}\int_0^{12} f(t)\cos\frac{\pi t}{3}\,dt \qquad \text{and} \qquad b_2 = \frac{1}{6}\int_0^{12} f(t)\sin\frac{\pi t}{3}\,dt.$$

4. For $f(t)$ as in the example, use the Trapezoidal Rule and Exercise 3 to estimate a_2 and b_2, the coefficients of $\cos \dfrac{\pi t}{3}$ and $\sin \dfrac{\pi t}{3}$, respectively.

5. Table 5.5 shows the average minimum temperatures for Boston, Massachusetts. We assume that

$$f(t) \approx a_0 + a_1 \cos \frac{\pi t}{3} + b_1 \sin \frac{\pi t}{3}$$

and

$$a_0 = \frac{1}{6}\int_0^{6} f(t)\,dt, \qquad a_1 = \frac{1}{3}\int_0^{6} f(t)\cos\frac{\pi t}{3}\,dt \qquad \text{and} \qquad b_1 = \frac{1}{3}\int_0^{6} f(t)\sin\frac{\pi t}{3}\,dt.$$

Use the Trapezoidal Rule for $n = 6$ to approximate a_0, a_1, and b_1. Note that these sines and cosines have period 6.

TABLE 5.5

January 22° F	July 64° F
March 30° F	September 56° F
May 49° F	November 37° F

6 INVERSE FUNCTIONS

APPLICATION 6.1: Human Volume

> See Section 6.1
> **The Logarithmic Function and Differentiation**
> **Calculus, 4th Edition,**
> **Larson/Hostetler/Edwards**

The relationships between the volume, weight, and the radius of a solid uniform ball is

$$W = V\rho$$

$$W = \left(\frac{4\pi}{3}\right) r^3 \rho$$

where ρ is the density of the ball, V is its volume, and r is its radius. But is this result, which is true for a ball, also valid for objects with other shapes? What about size and weight relationships for human beings?

EXAMPLE

Human Height and Weight

Can we apply our weight-volume model to human beings? Humans are not of uniform density, but we are close to uniform. We know that bones are more dense than soft tissue; however, for the purpose of this problem, we will assume that human beings have uniform density. Consider Table 6.1 which gives data for adults of medium frame.*

TABLE 6.1

Height (inches)	59	60	61	62	63	64	65	66
Male Weight (pounds)				128	131	135	139	142
Female Weight (pounds)	114	116	118	121	124	128	131	135

Height (inches)	67	68	69	70	71	72	73	74
Male Weight (pounds)	146	150	154	158	162	167	172	177
Female Weight (pounds)	139	142	146	150	153			

(a) Is there some simple relationship between weight and height for adult men of medium frame?

(b) If so, what is the rate of change of weight with respect to height?

---

*Source: *Family Health Guide and Medical Encyclopedia*, The Reader's Digest Association, Inc., Pleasantville, New York, 1970

SOLUTION

(a) If there is a simple power relationship between weight and height, then logarithms can be very helpful in studying it. We begin by considering the sphere once again. For a male human being, we will assume that there is a power relationship between height h and weight W, that is,

$$W = kh^n$$

or

$$\ln W = n \ln h + \ln k.$$

By least squares fitting* we determine that the slope of this curve appears to be $n = 1.84$ and $\ln k \approx -3.43$. Since we are only approximating, and our model is not perfect, the suggested value is $n = 2$ and we obtain

$$W = kh^2$$

where $k = 0.0324$ if h is in inches.

(b) We can now obtain the rate of change of W with respect to h. This will describe an incremental relationship between weight and height. Specifically, it will tell us, for example, approximately how much more a man should weigh, if he grows an inch taller. We differentiate the result from part (a)

$$W = 0.0324h^2$$

$$\frac{dW}{dh} = 0.0648h$$

This means for example, that for a man 66 inches tall, an increase in height of 1 inch should correspond to a $4\frac{1}{4}$-pound increase in weight. Try comparing the weights given by the model $W = 0.0324h^2$ with those given in Table 6.1.

---

* See Section 14.9, *Calculus*, 4th Edition, Larson/Hostetler/Edwards, 1990, D. C. Heath and Company, Lexington, MA.

APPLICATION 6.1 EXERCISES

1. How much would a 7-foot tall man of medium frame weigh?

2. If a 72-inch tall man shrinks $\frac{1}{2}$ inch per year during a particular year as he gets older, how much weight should he lose to stay at the average value for his new height?

3. Using the data from Table 6.1, graph accurately $\ln W$ versus $\ln h$ for the data for females. From the slope of this graph, verify that $n = 1\frac{2}{3}$ is reasonable. The appropriate k value is 0.129.

4. Suppose someone is shrinking an inch a day, such as in the science-fiction movie, "The Incredible Shrinking Woman." Find the rate at which her weight would be changing when she is 66 inches tall.

APPLICATION 6.2: Human Surface Area

> **See Section 6.1**
> **The Logarithmic Function and Differentiation**
> **Calculus, 4th Edition,**
> **Larson/Hostetler/Edwards**

A sky diver jumps from a plane. The drag on the diver's body is proportional to the diver's *surface area*. A medicine is applied to the skin. The correct dosage is proportional to the patient's surface area. For such cases as these, we might know the person's height or weight, but we must also know the person's surface area. We can model the relationship between height and surface area of a person by looking first at a simple, three-dimensional shape, the sphere. For a sphere, volume is

$$V = \frac{4}{3}\pi r^3$$

and surface area is

$$S = 4\pi r^2.$$

Surface area is thus the derivative of volume, with respect to the radius r. In two dimensions, for a circle, the boundary and the area are similarly related: $2\pi r$ is the derivative of πr^2. We might therefore expect that for people, the surface area (amount of skin) would be the derivative of volume. That is, if $V = kh^2$, then $S = 2kh$. Unfortunately, this model is not correct.

A much closer relationship, stated by Dubois and Dubois,* is

$$S = kh^{3/2}$$

where the constant $k = 1.43$ for people of medium build, h is measured in feet, and S is measured in square feet.

EXAMPLE

Height and Surface Area

At what rate does the surface area change with a person's height?

SOLUTION

We will use the Dubois and Dubois formula $S = kh^{3/2}$. By taking the derivative, we obtain

$$\frac{dS}{dh} = \frac{3}{2}kh^{1/2}.$$

Actual surface area depends on both height and weight. Our model works for people of medium build, but for people of slight or heavy build, the relationship can be further adjusted using the techniques of multivariable calculus.

**The Merck Manual, 14th Edition, Merck Sharp & Dohme Research Laboratories, 1982, p. 1836. The actual formula is $S = (\text{constant})(\text{weight})^{0.425}(\text{height})^{0.725}$.

APPLICATION 6.2 EXERCISES

1. If an adult male of medium build is 70 inches tall and he grows one additional inch, how much more skin (surface area) is added to his body?

2. Suppose the ideal dose of a particular drug for a 5-foot-tall person is 60 milligrams. The drug need is proportional to surface area. How much drug should be given to someone that is 6 feet tall, assuming medium build?

3. How much sunscreen would a person 5'10" use if a person 5'2" uses 10 milliliters?

4. Use the formula

$$S = (\text{constant})(\text{weight})^{0.425}(\text{height})^{0.725}$$

to develop a model for the surface area of a man of medium build.

APPLICATION 6.3: Friction and a Rod

> See Section 6.4
> **Exponential Functions:**
>   **Differentiation and Integration**
> **Calculus, 4th Edition,**
> **Larson/Hostetler/Edwards**

A cable is wrapped around a rod, as shown in Figure 6.1. Suppose we increase the friction by wrapping the cable around the rod again. What will happen to the tension in the rope? What will happen if we double the number of loops around the rod?

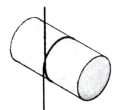

FIGURE 6.1

Our intuition in this case is guided by our knowledge of friction. We are usually taught in elementary physics that friction is independent (or nearly independent) of surface areas that are in contact with each other. However, in the case of wrapping a cable around a rod, we get unexpected results. Our first thought might be that the increase in friction and, therefore, the tension in the cable would be linear, that doubling the number of wraps would double the effects of friction. Not so! Here, instead of following our intuition, we should calculate mathematically the ratio of tensions in the cable.

EXAMPLE

Tug-of-War

Suppose a rope is wrapped about a tree trunk, perhaps in a game of tug-of-war. The tension in the ends of the rope is affected by the number of wrappings around the trunk. Use your knowledge of friction of a cable wrapped about a rod to see what effect the amount of wrapped cable has on the tug-of-war game.

SOLUTION

Consider a small segment of the cable or rope around the rod. (See Figure 6.2.) The normal force would be given by

$$F_n = 2T \sin\left(\frac{\theta}{2}\right).$$

For a small value of θ, the change in the force is

$$F_n = 2T\frac{\Delta\theta}{2} = T\Delta\theta.$$

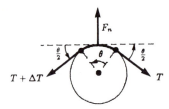

FIGURE 6.2

Assuming that the tension is less at one end of the cable, the tension will increase as

$$\Delta T = \mu F_n = \mu T \Delta \theta$$

where μ is the coefficient of friction. Rearranging terms and integrating produces

$$\frac{\Delta T}{T} = \mu \Delta \theta$$

$$\int \frac{dT}{T} = \int \mu \, d\theta$$

$$\ln T = \mu \theta + C.$$

Since $T = T_0$ when $\theta = 0$, we have

$$T = T_0 e^{\mu \theta}.$$

So the ratio of tensions from one end of the cable to the other is

$$\frac{T}{T_0} = e^{\mu \theta}.$$

With this ratio, the system will still be in static equilibrium. This explains why a few wraps of a rope around a tree makes for a very unfair game of tug-of-war!

APPLICATION 6.3 EXERCISES

1. A string is wrapped around a horizontally mounted wooden rod and weights are hung on each end until the string is about to slip. Keeping a constant weight $m_0 = 0.1$ kilogram on one end of the string, the following weights are found as the number of turns around the rod is varied.

Number of Turns, n	$\frac{1}{2}$	$1\frac{1}{2}$	$2\frac{1}{2}$	$3\frac{1}{2}$	$4\frac{1}{2}$	$5\frac{1}{2}$
Weight m kilograms	0.15	0.34	0.78	1.76	3.90	9.10

Remembering that each turn represents an angle of 2π radians, calculate the coefficient of friction between the string and the rod.

2. What weight could be supported if two more turns were added in Exercise 1 (assuming the string does not break)?

3. A stevedore is able to hold a ship still against a wind force of 6000 pounds by taking $5\frac{1}{4}$ turns of rope around a bollard on the dock and exerting a pull of 100 pounds on the free end of the rope. Calculate the coefficient of friction between the rope and the bollard.

APPLICATION 6.4: Measuring Earthquakes

> See Section 6.4
> Exponential Functions:
>   Differentiation and Integration
> Calculus, 4th Edition,
> Larson/Hostetler/Edwards

A scale for measuring the magnitude of earthquakes was developed in 1935 by Charles F. Richter of the California Institute of Technology. The so-called Richter Scale allows the "size" of earthquakes to be compared. The Richter formula computes the magnitude of a quake from the logarithm of the amplitude of waves recorded by seismographs. Included in the formula is an adjustment to compensate for the variation in the distance of the seismograph and the earthquake's epicenter. Some computed values for magnitude for some 20th century earthquakes are shown Table 6.2.

TABLE 6.2

Major Earthquakes		
Date	Place	Magnitude
1989	San Francisco	7.1
1985	Mexico City	8.1
1980	Italy	7.2
1978	Iran	7.7
1976	China	8.2
1971	San Fernando Valley	6.5
1964	Alaska	8.5
1933	Japan	8.9
1906	San Francisco	8.3

The Richter Scale formula for magnitude M is given by

$$M = \log_{10}\left(\frac{x}{c}\right)$$

where x is the amplitude of the largest seismic wave as measured 100 kilometers from the epicenter and c is the amplitude of a reference earthquake of amplitude 1 micron (1 micron = 0.001 mm) on a standard graph at the same distance from the epicenter.

EXAMPLE

San Francisco Earthquake of 1989

In 1989, San Francisco was struck by an earthquake of magnitude 7.1. As destructive as this earthquake was, it was not nearly as powerful as the 1906 San Francisco earthquake, which measure 8.3. The relative strengths of two earthquakes may be compared by looking at the ratio of the amplitudes. What is the ratio of amplitudes for these two San Francisco earthquakes?

SOLUTION

Using the Richter formula, set $M = 8.3$ for the 1906 earthquake.

$$8.3 = \log_{10}\left(\frac{x}{c}\right)$$
$$x = c10^{8.3}$$

Similarly, for the 1989 earthquake,

$$7.1 = \log_{10}\left(\frac{y}{c}\right)$$
$$y = c10^{7.1}.$$

Hence,

$$\frac{x}{y} = \frac{c10^{8.3}}{c10^{7.1}} = 10^{1.2} \approx 15.8.$$

The amplitudes of the two quakes are in the ratio of 15.8 to 1, so we can say that the earthquake in 1906 was 15.8 times as powerful as the earthquake in 1989.

APPLICATION 6.4 EXERCISES

1. The largest earthquake magnitude ever measured was 8.9 for an earthquake in Japan in 1933. The largest earthquake magnitude ever recorded in the United States was 8.5 for the Alaskan earthquake of 1964. Determine the ratio of amplitudes for these earthquakes.

2. When the amplitude of an earthquake is tripled, by how much does the magnitude increase?

3. Assume that there are two earthquakes of unequal strength. If the ratio of amplitudes is 1.5 and the weaker earthquake has magnitude 5.6, determine the magnitude of the stronger earthquake.

4. If the difference in magnitudes of two earthquakes is 0.5, determine the ratio of amplitudes for the two earthquakes.

5. Show that each whole number increase in magnitude represents a tenfold increase in measured amplitude.

APPLICATION 6.5: Measuring Sound Levels

> See Section 6.5
> **Bases Other Than e and Applications**
> **Calculus, 4th Edition,**
> **Larson/Hostetler/Edwards**

Sound is measured in decibels. One decibel (1 dB) is defined as

$$1 \text{ dB} = 10 \log_{10}\left(\frac{I}{I_0}\right)$$

where I_0 is the lowest intensity that humans can hear $\left(I_0 = 10^{-12} \text{ watts/meter}^2\right)$ and I is the sound intensity that is heard. Thus, it is the *logarithm* of an intensity ratio that gives the measure of loudness. Sound intensity depends on the distance from the sound source—it decreases as the reciprocal of the distance squared.

EXAMPLE

Sound of a Volcano

The volcano Krakatoa erupted on August 27, 1883 in what was possibly the greatest explosion on earth in modern times. On Rodriguez Island, 3000 miles away, the sound was reported to be as loud as the "roar of heavy guns." It was later estimated that the sound of the volcano erupting could be heard over $\frac{1}{13}$ of the surface of the earth, and had the earth been flat, and the sound not affected by the curvature of the earth, everyone on the earth would have heard it! What was the intensity of the sound 1 mile from Krakatoa?

SOLUTION

We will estimate the "roar of heavy guns" to be 100 decibels. This is a reasonable estimate since a rock concert would be 120 decibels and a power mower would be 100 decibels. Intensity decreases as the reciprocal of distance squared, so

$$\frac{I(1)}{I(3000)} = \frac{3000^2}{1^2}.$$

Our estimate for the sound at 3000 miles is 0.01 watts/meter^2. Therefore, the intensity at 1 mile was 90,000 watts/meter^2. The decibel level then is

$$\begin{aligned}
\text{dB level} &= 10 \log_{10}\left(\frac{I}{I_0}\right) \\
&= 10 \log_{10}\left(\frac{90,000}{10^{-12}}\right) \\
&= 170 \text{ dB}.
\end{aligned}$$

APPLICATION 6.5 EXERCISES

1. Suppose you are standing in the center of a large circle of young adults playing a game. Each person in the circle is screaming with happiness at 90 decibels (as measured one foot from each person). There are 100 people in the screaming circle, which has a radius of 30 feet. What is the intensity of the sound, in decibels, at the center of the circle?

2. Now imagine that you are in the middle of a very large crowd in which everyone is talking. Consider yourself isolated from the crowd by a circle of radius 1 foot. Imagine the effect of the talking crowd to be a two-dimensional uniform sound source of 40 decibels/square foot which extends outward from you in all directions for 100 feet. (Assume that the 40-decibel sound is measured one foot from the source.) Calculate, in decibels, the intensity of the sound that you would hear.

3. For the situation in Exercise 2, what would the intensity be if the crowd extended to infinity?

4. It is believed currently that any sound level over 90 decibels is harmful. Suppose the level of sound 30 feet from the sound system at an outdoor rock concert is 130 decibels. How far from the sound system should a person stand to avoid being harmed?

APPLICATION 6.6: Newton's Method

> **See Section 6.6**
> **Growth and Decay**
> **Calculus, 4th Edition,**
> **Larson/Hostetler/Edwards**

Frequently, remote areas are accessible only by air. But landing in such areas may be difficult due to weather conditions, the natural terrain, or civil disturbance. In such cases, food and provisions can be dropped from a plane.

The velocity, the height of the plane, the weight of the package, and the force due to air resistance all play a part in determining when the supplies reach the ground. We can describe these conditions using a differential equation, and by integration obtain the height of the package as a function of time t. Then by Newton's method we can find when the height is zero; that is, when the package hits the ground.

EXAMPLE

Delivering Rescue Provisions

A supply package weighing 160 pounds is dropped out of an airplane flying at an altitude of 5200 feet. For velocity v, the force due to air resistance is given by $-bv$, where b is a constant that depends on the size and nature of the body and on the viscosity of the air. In this example, we assume that $b = 1$. Use Newton's Method to determine the time at which the package reaches the ground.

SOLUTION

This problem appears to be straightforward, but it requires solving a transcendental equation! Let x represent the height of the object above the ground and t the time in seconds. First we will find velocity as a function of time. The net force

$$F = ma = m\frac{dv}{dt} = -160 - v.$$

The mass of the object is $m = \frac{160}{32} = 5$ slugs, and thus we must solve

$$5\frac{dv}{dt} = -160 - v.$$

We separate variables to obtain

$$\frac{dv}{160 + v} = -\frac{1}{5}\,dt.$$

Integration of both sides yields

$$\ln(160 + v) = -\frac{1}{5}t + C$$

and thus

$$160 + v = e^{-(1/5)t+C} = C_1 e^{-t/5}.$$

We use the fact that $v = 0$ when $t = 0$ to get $C_1 = 160$. So,

$$v = -160 + 160e^{-t/5}.$$

Now we will find x as a function of t.

$$v = \frac{dx}{dt} = -160 + 160e^{-t/5}$$

Integration yields

$$x = -160t - 800e^{-t/5} + C_2.$$

Since $x = 5200$ when $t = 0$, we obtain $C_2 = 6000$. Thus,

$$x = -160t - 800e^{-t/5} + 6000.$$

To see what value of t corresponds to the package hitting the ground, we set $x = 0$.

$$0 = -160t - 800e^{-t/5} + 6000$$

$$= -2t - 10e^{-t/5} + 75$$

To solve the equation, we let $f(t) = -2t - 10e^{-t/5} + 75$ and apply Newton's Method.

$$t_{n+1} = t_n - \frac{f(t_n)}{f'(t_n)}$$

$$f(t) = -2t - 10e^{-t/5} + 75$$

$$f'(t) = -2 + 2e^{-t/5}$$

Note that $f(35) > 0$ but $f(40) < 0$, so the root is between 35 and 40.

$$t_1 = 40$$

$$t_2 = 40 - \frac{f(40)}{f'(40)} = 37.497483$$

Continuing in this way, we obtain

$$t_3 = 37.497233$$

$$t_4 = 37.497233.$$

These two approximations are the same to 6 decimal places. Thus, the package hits ground in approximately 37.5 seconds.

APPLICATION 6.6 EXERCISES

1. Solve the example as given but change the weight of the object to 32 pounds and the height to 32 feet.
2. Solve the example as given but change the weight of the object to 64 pounds and the height to 32 feet.
3. Solve the example as given but assume that the force due to air resistance is given by $-2v$.
4. Solve the example as given but assume that it is a sky diver that falls and opens a parachute after 10 seconds. Assume that the air resistance is given by $-2v$ for time t greater than 10 seconds.

APPLICATION 6.7: Mixture Problems

> **See Section 6.6**
> **Growth and Decay**
> **Calculus, 4th Edition,**
> **Larson/Hostetler/Edwards**

Building ventilation problems have become a serious public concern. The Environmental Protection Agency estimates that as many as eight million homes in the United States have unsafe levels of radon.* By closing leaks and increasing insulation in buildings, we often sharply increase the concentration of pollutants in our environment.

   Here we shall consider the case of carbon dioxide, but the arguments used could be applied to other gases such as carbon monoxide and radon, which are far more dangerous. Carbon dioxide normally makes up 0.035% of the air we breathe but in rooms filled with people, this percentage could be considerably higher. Using calculus, we can model a gas-mixing situation and find the amount of CO_2 in a room after a brief time and after a steady state has been achieved.

EXAMPLE

Measuring Carbon Dioxide in a Room

A room of volume 2000 cubic feet initially contains 0.1% carbon dioxide. Starting at time $t = 0$, fresh air containing 0.035% carbon dioxide flows in at the rate of 100 cubic feet/minute. The well-mixed air in the room flows out at the same rate.

(a) Find the amount of carbon dioxide in the room after t minutes.

(b) Find the steady-state amount of carbon dioxide by letting $t \to \infty$.

SOLUTION

(a) Let x represent the amount of CO_2 in cubic feet in the room at time t. We set up the basic equation

$$\frac{dx}{dt} = \text{Rate In} - \text{Rate Out}$$

with the initial condition that at $t = 0$, $x = (0.001)(2000) = 2$ cubic feet. We begin by determining the Rate In.

$$\text{Rate In} = (0.00035)(100) = 0.035 = \frac{35}{1000} = \frac{7}{200} \text{ ft}^3/\text{min}$$

Next, we determine the Rate Out. We first find the concentration at time t, which is given by the relationship

$$\text{Concentration} = \frac{\text{Amount}}{\text{Volume}}.$$

Thus,

$$\text{Concentration of CO}_2 = \frac{\text{Amount of CO}_2}{\text{Volume}} = \frac{x}{2000}.$$

---

*See *Consumer Reports,* October, 1989.

Then,

$$\text{Rate Out} = (100)\left(\frac{x}{2000}\right) = \frac{x}{20}\ \text{ft}^3/\text{min}.$$

We now have the basic equation

$$\frac{dx}{dt} = \frac{7}{200} - \frac{x}{20} = \frac{7 - 10x}{200}.$$

Separating variables, we obtain

$$\frac{dx}{7 - 10x} = \frac{1}{200}dt.$$

Integration yields

$$-\frac{1}{10}\ln(7 - 10x) = \frac{1}{200}t + C_1$$

$$7 - 10x = e^{-(1/20)t - 10C_1} = e^{-10C_1}e^{-(1/20)t} = C_2 e^{-t/20}$$

Therefore,

$$x = 0.7 + 1.3e^{-t/20}.$$

This relationship gives the amount x in terms of time t.

(b) Letting $t \to \infty$, we see that $x \to 0.7$ cubic feet. Notice that eventually the concentration of CO_2 approaches the concentration of CO_2 in the incoming air.

APPLICATION 6.7 EXERCISES

1. At what time will the concentration of carbon dioxide be 0.05%?

2. One hundred gallons of salt water initially contain 20 pounds of dissolved salt. Starting at $t = 0$, salt water in a concentration of 3 pounds/gallon flows in at the rate of 4 gallons/minute. The well-stirred fluid flows out at the same rate.

(a) Find the amount of salt in the tank after t minutes.

(b) Find the steady-state amount of salt.

3. A lake of volume 100 cubic kilometers initially has a concentration of a particular pollutant of 0.01%. A polluted river which contains 0.05% of the pollutant flows in at the rate of 25 cubic kilometers/year.

(a) If the well-mixed fluid leaves at the same rate, find the concentration of the pollutant at time t.

(b) When will the concentration of the pollutant be twice the initial concentration?

4. *Blue baby* is the term used to describe a newborn with an Rh-blood-factor problem. Such infants must have their blood replaced, or at least partially replaced at birth, if they are to survive. In actual practice, 30 milliliters of blood is removed and then 30 milliliters of new blood is transfused. This process is repeated 15 times until one unit of blood (450 milliliters) is removed and the same amount of new blood is transfused. One unit of blood is nearly all the blood a newborn has! Find the final ratio of the remaining bad blood to the infant's total blood supply at the end of the procedure. *Hint:* To simplify this problem, assume that the blood is transfused continuously and that its mixing is instantaneous and complete.

APPLICATION 6.8: A Temperature Model

> See Section 6.8
> **Inverse Trigonometric Functions:**
>   **Integration and Completing the Square**
> **Calculus, 4th Edition,**
> **Larson/Hostetler/Edwards**

Temperature is a measurement of the average kinetic energy of molecules within matter. For a "naive classical" model, we have an ideal gas; we assume that the velocities of the gas molecules are equally represented up to some cut-off value, that is, they are uniformly distributed. In a "relativistic" model, we have maximum velocities that approach the speed of light, as for example, within the core of some stars.

To arrive at temperatures formed from the naive classical and relativistic models, we can calculate the *average* kinetic energies of the two models.

EXAMPLE

Classical and Relativistic Temperature

(a) In the naive classical model, use the uniformly distributed gas to determine T_c, the classical temperature for an ideal gas.

(b) In the relativistic model, we also assume a uniform distribution of gas molecules by velocity up to some cut-off value. The kinetic energy will be expressed as a relativistic quantity.

SOLUTION

(a) First we obtain the temperature in the naive classical gas model. In this model we will assume that the mass is constant, m_0, and that the maximum velocity is $0.9c$. Thus, the kinetic energy classically is

$$E_c = \frac{1}{2}m_0 v^2.$$

The naive classical temperature T_c is then proportional to E_c, so we have

$$T_c = \frac{K}{0.9c}\int_0^{0.9c} \frac{1}{2}mv^2\, dv$$

$$= \left(\frac{K}{0.9c}\right)\frac{m_0 v^3}{6}\Bigg]_0^{0.9c}$$

$$= Km_0 \frac{0.81c^2}{6}$$

$$= 0.135 Km_0 c^2.$$

These models are actually flawed because the velocities of gas molecules are not uniformly distributed as we assumed in our models. Instead, the number of molecules at any given speed is expressed by Maxwell's Distribution

$$f(v) = 4\pi N\left(\frac{m}{2\pi kT}\right)^{3/2} v^2 e^{-(mv^2)/(2kT)}$$

where m is the mass of each molecule, k is Boltzman's constant, and T is the temperature in Kelvin. This formula was developed by James Clerk Maxwell (1831–1897). To obtain the classical average kinetic energy we integrate using Maxwell's Distribution function.

(b) For the relativistic model, we begin with the relativistic kinetic energy E_r expressed as

$$E_r = m_0 c^2 \left[\frac{1}{\sqrt{1 - (v/c)^2}} - 1 \right].$$

The relativistic temperature T_r is given by

$$T_r = \left(\frac{K}{0.9c} \right) \int_0^{0.9c} m_0 c^2 \left[\frac{1}{\sqrt{1 - (v/c)^2}} - 1 \right] dv$$

$$= \left(\frac{K}{0.9c} \right) m_0 c^3 \left[\arcsin \left(\frac{v}{c} \right) - \left(\frac{v}{c} \right) \right]_0^{0.9c}$$

$$= \left(\frac{K}{0.9c} \right) m_0 c^3 [\arcsin(0.9) - 0.9]$$

$$\approx \left(\frac{K m_0 c^3}{0.9} \right) (1.1197 - 0.9)$$

$$\approx 0.244 K m_0 c^2.$$

APPLICATION 6.8 EXERCISES

1. Obtain the classical value for the average kinetic energy for an ideal gas which satisfies Maxwell's Distribution Law. In the classical case, $0 \leq v < \infty$.
2. For a cut-off speed of $0.5c$, compare T_c and T_r.

7

APPLICATIONS OF INTEGRATION

74   CHAPTER 7

APPLICATION 7.1: Work and Heat Engines

> See Section 7.1
> Area of a Region Between Two Curves
> Calculus, 4th Edition,
> Larson/Hostetler/Edwards

Thermodynamics allows us to study ideal models for a heat engine that converts heat energy to mechanical energy. For an idealized engine, we will assume the processes to be pure and that gases obey ideal gas laws such as $PV = nRT$, or for adiabatic expansions, $PV^\gamma = $ constant ($\gamma \approx 1.4$ for air). We will examine the work that such an idealized engine could perform.

The work done by the system in one complete cycle in an ideal heat engine is given by the area enclosed in the cycle when illustrated on a P-V (pressure versus volume) diagram. This can be seen by considering the definition of work.

$$\text{Work} = \int_a^b F\,dx = \int_a^b PA\,dx$$

where P is pressure and A is a cross-sectional area. But $A\,dx = dV$, so

$$\text{Work} = \int_{V_a}^{V_b} P\,dV.$$

Net work for an entire cycle would therefore be the area enclosed within the region on the P-V diagram.

EXAMPLE

An Ideal Steam Engine: The Rankine Cycle

An example of a heat engine is the Rankine cycle, which is the ideal version of a steam engine. (See Figure 7.1.)

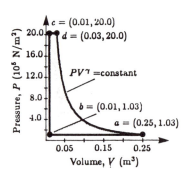

FIGURE 7.1

Liquid water at low temperature and pressure is heated at constant volume within a boiler. This occurs along path bc. The water is converted to steam and expands along path cd. The expansion continues adiabatically along path da. The water is then cooled and condensed back to liquid along path ab.

There is no work done along path bc because the volume remains constant. From c to d the work is $P_c(V_d - V_c)$, which is 0.4×10^5 joules. The steam expands adiabatically along da. The law for such expansions is $PV^\gamma = $ constant. The resulting expression for the work is

$$\text{Work (adiabatic path)} = \frac{P_a V_a - P_d V_d}{1 - \gamma}.$$

Substituting in the values, we find that the work along path da is 0.85625×10^5 joules.

Work in returning from point a to b is negative and is similar to the evaluation for path cd. The work along this path is -0.2472×10^5 joules. The net work is $(0.4 + 0.85625 - 0.272) \times 10^5$ joules.

APPLICATION 7.1 EXERCISES

1. Derive the expression for work along an isothermal path.

$$\text{Work from } a \text{ to } b = nRT \ln\left(\frac{V_b}{V_a}\right)$$

2. Derive the expression for work along an adiabatic path.

$$\text{Work from } a \text{ to } b = \frac{P_b V_b - P_a V_a}{1 - \gamma}$$

3. The gasoline engine is modeled by the Otto cycle. (See Figure for Exercise 3.) Air and gasoline are placed in the cylinder at point b. The mixture is compressed adiabatically to point c. A spark is provided and the pressure increases very rapidly to point d. The hot gas then expands adiabatically to point a, where the combustion products are exhausted. Calculate the net work in one cycle of this engine.

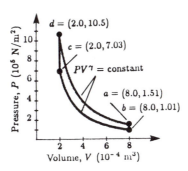

FIGURE FOR EXERCISE 3

4. The Diesel cycle (See Figure for Exercise 4) begins with compressing air at point b along the path to point c. This compression is large enough to make the interior of the cylinder very hot. A slow-burning fuel is sprayed into the cylinder at point c. As it burns slowly, it expands the volume to point d. The combustion products continue to expand adiabatically to point a, where the combustion products are exhausted and the cycle repeats. Calculate the net work in this cycle.

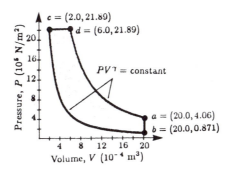

FIGURE FOR EXERCISE 4

5. The Brayton cycle heat engine (see Figure for Exercise 5) is the model for a gas turbine engine. Air is drawn into the compressor and is compressed adiabatically along path *bc*. Fuel is mixed and burned along path *cd*. The burnt fuel mixture continues to expand adiabatically along path *da*. The combustion chamber is cleared of exhaust along *ab*. Find the net work for one cycle of the Brayton cycle.

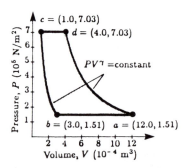

FIGURE FOR EXERCISE 5

APPLICATION 7.2: Power Distribution

> **See Section 7.1**
> **Area of a Region Between Two Curves**
> **Calculus, 4th Edition,**
> **Larson/Hostetler/Edwards**

In large power-generating systems, many generators are connected together through cables. (Such a system is called a bus.) An example of such a system is Pacific Intertie; it connects generating facilities such as Pacific Gas and the Electric Helms Facility in a power generating network. The output of each generator must be at the correct frequency (60 Hertz in the United States). In addition, the generator output must be in phase, or nearly in phase, with the bus. If the generator output was at the correct frequency but out of phase, it would actually reduce the total available electricity because the out-of-phase generator would be canceling the output of other generators.

Suppose we take as our model for power distribution a single synchronous generator connected to an infinite bus—a good model for the Pacific Intertie, where there are so many generators connected to the facility that it appears to be a single generator and a nearly infinite bus.

EXAMPLE

Stability of a Generator

For a large power-generating system, under what conditions is the system stable? Using both a graph of power angle versus input power and the following differential equation, we can develop a criterion for system stability. In a normalized set of units, the differential equation that describes the system is

$$\frac{2H}{\omega} \frac{d^2\delta}{dt^2} = P_m - P_e$$

where H is a constant, ω is the angular speed of the generator, P_m is the mechanical input power, and P_e is the output power.

We define the power angle δ as the angle that the output power lags behind the input power due to load. The output power is a function of δ and the maximum input power $P_{\max}$.

$$P_e = P_{\max} \sin \delta$$

(a) Derive an expression for stability of the system called the "equal areas criterion," which states that the system will be stable if the shaded areas on the graph of power angle versus power input are equal. See Figure 7.2.

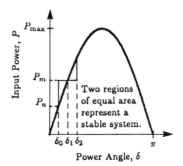

FIGURE 7.2

(b) If P_m represents the mechanical input power, P_m does not vary with respect to δ because δ is a measurement of the lag in power. P_m can change for such reasons as a sudden increase in torque to the generating turbines. Suppose there is a sudden increase in the input power from P_m to P_n. Under what condition will this produce a stable equilibrium situation?

SOLUTION

(a) We recognize that the differential equation is essentially the same second-order differential equation that Newton's Law gives for a pendulum.

$$ma = -mg\sin\theta$$

$$\frac{d^2\theta}{dt^2} = -\frac{g}{L}\sin\theta$$

Just as with Newton's equation for small θ, or for small δ in our example, we will get simple harmonic motion. We will obtain an alternative solution without using the small oscillation approach. Note that by the Chain Rule we obtain

$$\frac{d}{dt}\left[\frac{d\delta}{dt}\right]^2 = 2\frac{d\delta}{dt}\frac{d^2\delta}{dt^2}.$$

We multiply by H/ω and rewrite.

$$\frac{H}{\omega}\frac{d}{dt}\left[\frac{d\delta}{dt}\right]^2 = \frac{2H}{\omega}\frac{d^2\delta}{dt^2}\frac{d\delta}{dt}$$

$$= (P_m - P_e)\frac{d\delta}{dt}$$

Integrating with respect to t from δ_0 to δ_2 produces

$$\frac{H}{\omega}\int_{\delta_0}^{\delta_2}\frac{d}{dt}\left[\frac{d\delta}{dt}\right]^2 dt = \frac{H}{\omega}\left(\frac{d\delta}{dt}\right)^2\Bigg]_{\delta_0}^{\delta_2}$$

$$= \int_{\delta_0}^{\delta_2}(P_m - P_e)\,d\delta.$$

For simple harmonic motion or other similar behavior, maximum displacement will correspond to zero velocity, so where $\delta = \delta_0$ and $\delta = \delta_2$ we obtain

$$\frac{d\delta}{dt} = 0.$$

This implies that

$$0 = \int_{\delta_0}^{\delta_2}(P_m - P_e)\,d\delta = \int_{\delta_0}^{\delta_1}(P_m - P_e)\,d\delta + \int_{\delta_1}^{\delta_2}(P_m - P_e)\,d\delta$$

$$\int_{\delta_0}^{\delta_1}(P_m - P_e)\,d\delta = \int_{\delta_1}^{\delta_2}(P_e - P_m)\,d\delta.$$

This is a statement that stability occurs when the shaded areas shown in Figure 7.2 are equal. We have thus derived an equal-areas criterion for stability.

(b) If the load is suddenly increased, the power angle is shifted to δ_1. The system will oscillate between δ_0 and δ_2 if stability is to be achieved. See Figure 7.3. Once again the shaded areas must be equal.

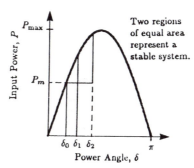

FIGURE 7.3

APPLICATION 7.2 EXERCISES

1. Assume initially that the power angle δ is $\pi/6$ and the new equilibrium is shifted to $\pi/3$. Find δ_2.

2. If $\delta_{\max}$ is larger than $\pi - \delta_0$, then no stability is possible since the areas cannot be equal. δ_{critical} is the largest change in the equilibrium power angle such that stability is possible. It is clear from Figure 7.3 that when $\delta_2 = \pi - \delta_0$ the generator is about to become unstable. Find δ_{critical} if $\delta_0 = \pi/6$, where the generator is about to become unstable.

APPLICATION 7.3: Archimedes Principle

> See Section 7.2
> Volume: The Disc Method
> Calculus, 4th Edition,
> Larson/Hostetler/Edwards

The Archimedes Principle states that the upward or buoyant force on an object within a fluid is equal to the weight of the fluid that the object displaced. Recall that weight is the special name given to the gravitational force an object experiences near the surface of the earth. There are two cases to consider when calculating this force. One case is when the object is entirely submerged. The other, more interesting case, is when the object is floating, that is, only partially submerged.

Consider the case of the partially submerged object. We can obtain information about the relative densities of the floating object and the fluid by observing how much of the object is above the surface of the fluid and how much is below the surface of the fluid. We can also determine how large a floating object is if we know the amount that is above the surface and we know the relative densities.

In the first reported practical use the Archimedes Principle, Archimedes was asked to determine whether a crown was real gold or a fake. This is not a trivial problem because the crown was not a simple shape. Archimedes thought for a long time before he came up with his ingenious solution. He weighed the crown in and out of water and from that, was able to conclude that the crown was fake!

EXAMPLE

Tip of the Iceberg

Suppose we can see the top of a floating iceberg. We know the density of ice is 0.92×10^3 kg/m^3 and of ocean water is 1.03×10^3 kg/m^3. Using Archimedes Principle, what is the total size of the iceberg?

SOLUTION

Consider some object of uniform density floating in a fluid. (See Figure 7.4.)

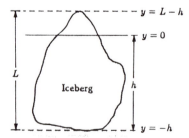

FIGURE 7.4

Let ρ_f be the density of the fluid and let ρ_o be the density of the object. The buoyant force is given by

$$F = \rho_f g \int_{-h}^{0} A(y)\, dy$$

where $A(y)$ is a typical cross section and g is the acceleration due to gravity. The weight of the object is

$$W = \rho_o g \int_{-h}^{L-h} A(y)\,dy.$$

We also know by the Archimedes Principle that when the object is floating $F = W$, so

$$\rho_f g \int_{-h}^{0} A(y)\,dy = \rho_o g \int_{-h}^{L-h} A(y)\,dy.$$

This gives the ratio of densities

$$\frac{\rho_o}{\rho_f} = \frac{\text{submerged volume}}{\text{total volume}}.$$

Since, for ice, the density is 0.92×10^3 kilograms/cubic meter and for ocean water, the density is 1.03×10^3 kilograms/cubic meter, about 89% of the iceberg is below the surface.

APPLICATION 7.3 EXERCISES

1. For totally submerged objects, obtain

$$\frac{\rho_c}{\rho_f} = \frac{w - W}{w}$$

where ρ_c is the density of a submerged object such as a crown, ρ_f is the density of the fluid, w is the weight in air, and W is the weight in the fluid.

2. Show that if a floating object has a constant horizontal cross section, a vertical displacement results in a restoring force proportional to the displacement. This sort of situation would result in simple harmonic motion if the fluid did not dampen the motion.

3. A ball of radius 0.1 meters is floating in a pool with $\frac{3}{4}$ of the vertical diameter above water. (The density of water is 1000 kg/m^3.) The ball is then forced down so that it is half submerged. Find the restoring force as described in Exercise 2.

APPLICATION 7.4: Moment of Inertia for Wheels

> See Section **7.3**
> Volume: **The Shell Method**
> Calculus, **4th Edition,**
> Larson/Hostetler/Edwards

In a rotating wheel, the energy due to rotation can be expressed in terms of the moment of inertia I. The energy is $E = \frac{1}{2}I\omega^2$, where ω is the angular velocity. It has been said for bicycles that removing a pound of weight from the wheels is like removing two pounds from elsewhere on the bicycle. This is only partially true because the wheels are both rotating and translating. This means that there are two terms to consider for kinetic energy: the translational term, $\frac{1}{2}mv^2$, and the rotational term, $\frac{1}{2}I\omega^2$. Recall that $\omega r = v$ along the outer edge of the wheel.

EXAMPLE

Energy in an Ideal Bicycle Wheel

We will try to model an ideal bicycle wheel. To make our model very simple, we will assume that the hub and spokes are massless. Since the hub is close to the axis of rotation, it will not contribute much to the moment of inertia. (Ignoring the spokes is not as easy to justify. We leave the justification to you as an exercise.) To make the model simpler, we will assume that the cross section of the rim, tire, and inner tube form a rectangle 3 cm by 4 cm, that the mass is uniform around the rectangle, and that the inner radius of the rim is 31 cm and outer radius is 35 cm, as shown in Figure 7.5. What is the ratio of kinetic rotational energy to kinetic translational energy for such an ideally modeled bicycle wheel?

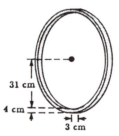

FIGURE 7.5

SOLUTION

The moment of inertia for the tire/rim assembly is given by three terms. The moment of inertia for the inner band I_a is

$$I_a = 2\rho\pi aLa^2 = 178,746\rho\pi$$

where a is the inner radius, ρ is the density of the wheel, L is the width, and for the outer band I_b is

$$I_b = 2\rho\pi bLb^2 = 257,250\rho\pi$$

where b is the radius. The moment of inertia for each of the sides I_s is

$$I_s = \frac{1}{2}\rho\pi(b^4 - a^4) = 288,551\rho\pi.$$

We can now write the ratio of rotational to translational energy for this simple model of a bicycle wheel.

$$\frac{E_R}{E_T} = \frac{\frac{1}{2}(I_a + I_b + 2I_s)\omega^2}{\frac{1}{2}m\omega^2 r^2}.$$

The mass of the wheel is

$$\rho\pi(b - a + L)(a + b) = 924\rho\pi.$$

Substituting all these values into the given equation, we get

$$\frac{1,013,100\rho\pi}{924 \times 35^2\rho\pi} = 0.895.$$

APPLICATION 7.4 EXERCISES

1. Determine the maximum possible ratio between the rotational and translational kinetic energy for an ideal wheel.

2. Determine the moment of inertia for a more realistic bicycle wheel in which the rim mass is 450 grams and the mass of each of the 36 spokes is 9 grams. Assume the inner tube mass is 140 grams; the tire mass is 450 grams. Treat the cross section of each item as a line of appropriate shape and uniform density except the hub (mass = 400 grams) which can be considered to be a hollow cylinder. (See figure for Exercise 2.)

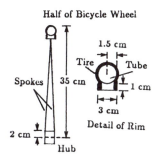

FIGURE FOR EXERCISE 2

3. Using the moment of inertia from Exercise 2, compute the ratio of rotational and translational energy for this wheel.

4. Determine the moment of inertia for a train wheel as illustrated in the figure for rotation through an axle. The density of steel is 7.8×10^3 kilograms/cubic meter.

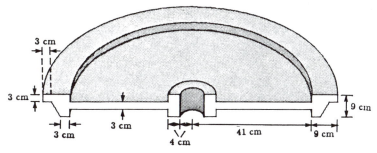

FIGURE FOR EXERCISE 4

APPLICATION 7.5: Hyperbolic Functions

> **See Section 7.4**
> **Arc Length and Surfaces of Revolution**
> **Calculus, 4th Edition,**
> **Larson/Hostetler/Edwards**

EXAMPLE

Hyperbolic Cosine Function

The hyperbolic cosine function has the property that the area under the curve equals the arc length. Let f be the function with this property

$$\int_0^x f(t)\, dt = \int_0^x \sqrt{1 + [f'(t)]^2}\, dt.$$

Show that the function f is the hyperbolic cosine function.

SOLUTION

Rewriting gives

$$\int_0^x f(t) - \sqrt{1 + [f'(t)]^2}\, dt = 0.$$

Since this is true for all x, the integrand must be zero, so

$$f(t) = \sqrt{1 + [f'(t)]^2}$$

$$[f(t)]^2 = 1 + [f'(t)]^2$$

$$[f'(t)]^2 = [f(t)]^2 - 1$$

$$\left[\frac{du}{dt}\right]^2 = u^2 - 1$$

$$\frac{du}{dt} = \sqrt{u^2 - 1}$$

$$\int \frac{du}{\sqrt{u^2 - 1}} = \int dt$$

$$\cosh^{-1} u = t + c$$

$$f(t) = \cosh(t + c).$$

See Figure 7.6.

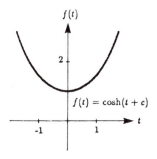

$f(t) = \cosh(t + c)$

FIGURE 7.6

APPLICATION 7.5 EXERCISES

1. Find the length of the graph of

$$f(x) = 20 \cosh \frac{x}{20}, \qquad -20 \ln 20 \le x \le 10 \ln 20.$$

2. Find the area of the region lying between the curve in Exercise 1 and the x-axis.

APPLICATION 7.6: The Big Egg

> **See Section 7.4**
> **Arc Length and Surfaces of Revolution**
> **Calculus, 4th Edition,**
> **Larson/Hostetler/Edwards**

A large egg was erected in Vegreville, Alberta, Canada, as the town monument. This egg monument has two meanings. First, it honors the 2000-year-old Ukrainian tradition of painting very intricate Easter eggs; and second, it recognizes the 200 years of peace and security provided by the Mounties to the Ukrainians in Vegreville. This particular egg has an aluminum frame and is tiled; however, it could have been made with steel mesh and concrete.

Eggs come in all shapes and sizes. However, for our model we will consider the egg to be the curve

$$r = R(2 + \cos\theta) = 5(2 + \cos\theta), \quad 0 \le \theta \le \pi, \quad R = 5$$

revolved about the x-axis. If r is in feet, this will produce an egg which is 20 feet long. (See Figure 7.7.)

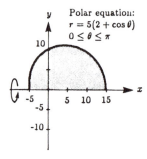

FIGURE 7.7

EXAMPLE

Building the Big Egg

How many cubic yards of material are needed if the eggshell is to be 0.2 feet thick? (Assume the material is concrete, which is measured in cubic yards.)

SOLUTION

There are several ways to obtain our desired result. We could calculate the volume of a solid egg with $R = 4.8$ feet and another with $R = 5.0$ feet and take the difference. Instead, we will calculate the volume as a function of the radius and use the differential to estimate the needed concrete. The volume integral is

$$V = \int_{-5}^{15} \pi y^2 \, dx$$

where $x = r\cos\theta$, $y = r\sin\theta$, and $dx = -r\sin\theta \, d\theta$. We then get

$$V = \pi \int_0^\pi r^3 \sin^3 \theta \, d\theta$$

$$= \pi \int_0^\pi [R(2 + \cos \theta)]^3 \sin^3 \theta \, d\theta$$

$$= \pi R^3 \int_0^\pi (8 + 12 \cos \theta + 6 \cos^2 \theta + \cos^3 \theta)(1 - \cos^2 \theta) \sin \theta \, d\theta$$

$$= \pi R^3 \int_0^\pi [8 + 12 \cos \theta - 2 \cos^2 \theta - 11 \cos^3 \theta - 6 \cos^4 \theta - \cos^5 \theta] \sin \theta \, d\theta.$$

We now let $u = \cos \theta$; this also changes the limits of integration from $0 \le \theta \le \pi$ to $1 \ge u \ge -1$. The resulting integral is

$$V = \pi R^3 \int_1^{-1} (-8 - 12u + 2u^2 + 11u^3 + 6u^4 + u^5) \, du.$$

By symmetry, all the odd terms will contribute nothing and we can integrate from 0 to 1 and double the results obtained from the even terms. Then

$$V = 2\pi R^3 \left[-8u + \frac{2u^3}{3} + \frac{6u^5}{5} \right]_0^{-1}$$

$$= \frac{184 \pi R^3}{15}.$$

Now we use the differential to estimate the volume and we get

$$dV = 3R^2 \left(\frac{184}{15} \right) \pi \, dR$$

$$= 36.8 \pi R^2 \, dR$$

$$\approx 578 \text{ ft}^3 \approx 21.4 \text{ cubic yards.}$$

So the big egg will require 21.4 cubic yards of material.

APPLICATION 7.6 EXERCISES

1. Calculate the surface area of the big egg and therefore, the amount of steel mesh needed to construct the egg.
2. Calculate the volume of the shell of the big egg if $4.9 \le R \le 5.1$ by actually calculating the volumes of solid eggs and then taking the difference in the volumes. Convert the result, which will be in cubic feet, to a value in cubic yards.
3. Calculate the material needed to construct a concrete eggshell 0.2 feet thick if the equation that represents the egg is $r = 4(3 + \cos \theta)$.
4. Obtain the number of square feet of steel mesh needed for reinforcement for the egg in Exercise 3.

APPLICATION 7.7: The Helms Pump Storage Facility

> See Section 7.5
> **Work**
> Calculus, 4th Edition,
> Larson/Hostetler/Edwards

High in the Sierra Nevada mountains, east of Fresno, California, is the Helms Pump Storage Facility.* This large power-generation plant, which has a total capacity of 1200 megawatts, is owned by Pacific Gas and Electric Company. One of a number of very interesting features of this facility is that it uses more electricity than it produces and yet operates cost effectively.

The design of the Helms facility allows it to pump water from the lower reservoir, Wishon, to the upper reservoir, Courtright, (see Figure 7.8), by using excess electricity that is produced at other generating plants. Thus, when the demand exceeds the output of the other generating plants on the electric grid, the Helms facility can reproduce its "stored electricity" by allowing the water from Courtright to flow back into Wishon through the facility's three generators.

The elevation of the upper reservoir, Courtright, is 8250 feet, and the elevation of the lower reservoir, Wishon, is 6500 feet. When the facility is in operation, the turbines that power the pumps are 1900 feet below Courtright. The expended water is dumped into Wishon well below the surface. This is done so that when the water levels are low, pumping from Wishon to Courtright will still be possible.

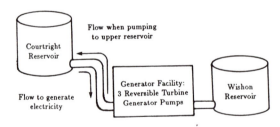

FIGURE 7.8

EXAMPLE

Calculating Work

In this example, we will calculate the work necessary to pump from a lower reservoir to an upper reservoir. For simplicity, each reservoir will be modeled as a cylinder, with the lower reservoir having a constant cross section of 800 million square feet and upper reservoir having a constant cross section of 1 billion square feet. The difference in height between the intake and the discharge tubes is assumed to be 1600 feet. For this calculation we assume that the lower reservoir has an initial water level of $y_1 = 75$ feet and the upper reservoir has an initial water level of $Y_1 = 100$ feet, as shown in Figure 7.9. Calculate the work necessary to pump water in the lower reservoir to the upper reservoir so that the depth of the lower reservoir will drop 1 foot.

---

*For more information see *The Hidden Power Plant,* Pacific Gas and Electric Company.

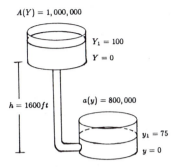

FIGURE 7.9

SOLUTION

The problem is easier to solve if it is broken into three parts: (a) pumping water from the lower reservoir to the bottom of the upper reservoir; (b) pumping water from the bottom of the upper reservoir to raise the level of the upper reservoir; and (c) combining parts (a) and (b) to obtain the desired solution.

(a) An increment of work for pumping water from the lower reservoir to the bottom of the upper reservoir is given by

$$\Delta W_1 = \rho a(y)(h - y)\,\Delta y$$

where $a(y)$ is the area of the cross section at a particular y, ρ is the density of water, and h is the difference in elevation between the pipe openings in both reservoirs. Note that y is zero at the pipe opening near the bottom of the reservoir.

(b) An increment of work for pumping from the bottom of the upper reservoir to raise its level is given by

$$\Delta W_2 = \rho A(Y)Y\,\Delta Y$$

where $A(Y)$ is the appropriate cross section and Y is measured from the pipe opening in the upper reservoir.

(c) The total work required to pump water from the lower reservoir to the upper reservoir is given by

$$W = \int_{y_1-1}^{y_1} \rho a(y)(h-y)\,dy + \int_{Y_1}^{Y_2} \rho A(Y)Y\,dY.$$

The decrease in volume in the lower reservoir must be the same as the increase in the volume of the upper reservoir. In this model, the cross-sectional areas are assumed to be constant. Therefore, the ratio of the cross-sectional areas is the reciprocal of the ratio of heights and $Y_2 = 100.8$ feet. Substituting the values of y_1, Y_1, and Y_2 into the integrals, we obtain

$$W = \int_{74}^{75} 0.8 \times 10^8 \rho(1600 - y)\,dy + \int_{100}^{100.8} 10^8 \rho Y\,dY$$

$$= \left[0.8 \times 10^8 \rho \left(1600y - \frac{y^2}{2} \right) \right]_{74}^{75} + \left[10^8 \rho \frac{Y^2}{2} \right]_{100}^{100.8}$$

$$= 10^8 \rho (1220.4 + 80.32)$$

$$= 1300.72\rho \times 10^8$$

$$\approx 1.30\rho \times 10^{11} \qquad\qquad (\rho \approx 62.4\ \text{lb/ft}^3)$$

$$\approx 8.11 \times 10^{12}\ \text{ft-lbs}$$

$$\approx 1.1 \times 10^{13}\ \text{joules}.$$

It is estimated that it would take three hours to pump this amount of water to the upper reservoir. This amount of water, when returned to the lower reservoir, would produce electricity at full capacity (1200 megawatts) for 2.5 hours!

APPLICATION 7.7 EXERCISES

1. Using the example as a model, calculate the amount of work required to pump water from a lower reservoir to an upper reservoir so that the depth of the lower reservoir drops by 2 feet. Assume, as in the example, that both reservoirs are cylindrical. The area of a cross section of the lower reservoir is 500,000 square feet and its water level is 60 feet. The area of a cross section of the upper reservoir is 750,000 square feet and its water level is 80 feet. The difference in height between the intake and discharge tubes is 1800 feet.

2. Water is stored in a cylindrical tank. The axis of the tank is horizontal, as shown in the figure. Find the work required to pump half of the water in the full (lower) tank into the (upper) spherical tank. Can more than half of the water in the lower tank be pumped into the upper tank?

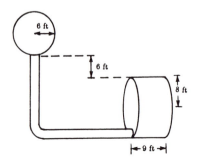

FIGURE FOR EXERCISE 2

3. Oil, that has a density of 50 lbs/ft^3, is pumped into the bottom of a closed, insulated chamber. The atmospheric pressure in the chamber is initially 14.7 lbs/in^2. The chamber has a height of 10 feet and a cross section of 1 square foot, as shown in the figure. Find the work needed to fill the chamber to a height of 9 feet. Assume the gas compresses adiabatically so it obeys the law $PV^\gamma =$ constant. The constant γ is called the *heat capacity ratio*. (For this problem, use $\gamma \approx 1.4$.) Recall that work done on a gas is $W = \int P \, dV$.

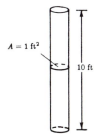

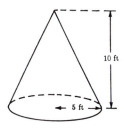

FIGURE FOR EXERCISE 3   FIGURE FOR EXERCISE 4

4. Repeat Exercise 3 for a conical chamber whose vertex is up, as shown in the figure. The height of the cone is 10 feet and the radius of the base is 5 feet.

APPLICATION 7.8: Climbing El Capitan

> See Section 7.5
> **Work**
> **Calculus, 4th Edition,**
> **Larson/Hostetler/Edwards**

The scaling of the nearly vertical 3569-foot El Capitan Mountain in Yosemite National Park would be both a physical and symbolic achievement for any climber. It was a particularly noteworthy achievement for Mark Wellman who is the first climber to scale El Capitan using only his arms. Despite being a paraplegic, Wellman reached the summit in a climb during the summer of 1989.

EXAMPLE

The Work of Mountain Climbing

Let's determine the work done by a climber using a rope to reach a summit. Suppose we consider a 160-pound mountain climber as he prepares for the final part of his climb. He is standing on a ledge 100 feet from the top of the mountain. (See Figure 7.10(a).) One end of a rope, which is 120 feet long, is secured to the top of the mountain. For the sake of safety, the other end is attached to the climber's chest. Assume the rope weighs $\frac{1}{4}$ pound per linear foot and is currently coiled at the feet of the climber. What is the amount of work done by the man in climbing up the rope to the top of the mountain? Assume that the climber's hands are at chest level.

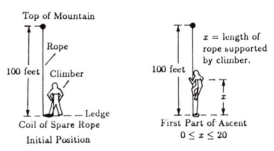

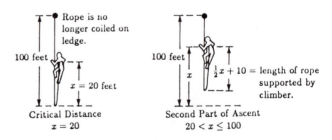

FIGURE 7.10

SOLUTION

Let x represent the height of the climber's hands above the ledge. The climber must lift the weight of the rope as well as his own height. The formula for the weight of the rope will change for $x > 20$, since from this point on, the rope will no longer be coiled on the ledge; the entire length of the rope will be suspended. (See Figure 7.10(c).) We will solve the problem in two parts. For $0 \le x \le 20$, the weight of the rope that is supported by the climber is given by

$$(\text{length})(\text{density}) = (x)\left(\frac{1}{4}\right).$$

This is because that part of the rope that is coiled on the ledge is not being supported by the climber. (See Figure 7.10(b).) Thus, the total weight is $160 + (x/4)$. So,

$$\text{Work } 1 = \int_a^b F(x)\,dx$$

$$= \int_0^{20} \left(160 + \frac{x}{4}\right) dx$$

$$= \left[160x + \frac{x^2}{8}\right]_0^{20}$$

$$= 3250 \text{ ft} \cdot \text{lb}.$$

For $20 \le x \le 100$, we need a formula for the length of the rope supported by the climber as a function of x. (See Figure 7.10(d).) The function is linear and satisfies the following values.

$$x = 20 \qquad y = 20$$
$$x = 100 \qquad y = \frac{120}{2} = 60$$

Therefore, the equation for this length is given by

$$\frac{y - 20}{x - 20} = \frac{60 - 20}{100 - 20} = \frac{1}{2}$$

$$y = \frac{1}{2}x + 10.$$

Hence, the weight of the rope supported by the climber is given by

$$(\text{length})(\text{density}) = \left(\frac{1}{2}x + 10\right)\left(\frac{1}{4}\right)$$

$$\text{total weight} = 160 + \left(\frac{1}{2}x + 10\right)\left(\frac{1}{4}\right)$$

$$= \frac{325}{2} + \frac{x}{8}.$$

This implies that

$$\text{Work } 2 = \int_{20}^{100} \left(\frac{325}{2} + \frac{x}{8} \right) \, dx$$

$$= \left[\frac{325}{2} x + \frac{x^2}{16} \right]_{20}^{100}$$

$$= 13,600 \text{ ft} \cdot \text{lb}.$$

Thus, the total work is given by

$$\text{Work } 1 + \text{Work } 2 = 3250 + 13,600 = 16,850 \text{ ft} \cdot \text{lb}.$$

APPLICATION 7.8 EXERCISES

1. Solve the example, except assume that the length of the rope is 100 feet.

2. Solve the example, except assume that the length of the rope is 200 feet.

3. Solve the example, except assume that the rope is initially hanging off the ledge. (See Figure for Exercise 3.)

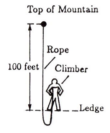

FIGURE FOR EXERCISE 3

4. Solve the example, except assume that the man weighs 180 pounds, the rope is 140 feet long, and the rope weighs $\frac{3}{8}$ pound per linear foot.

5. Actual mountain climbing rope comes in 165 foot lengths and has a total weight of 7 pounds. Solve the example using these measurements.

APPLICATION 7.9: Work Involving a Curved Path

> **See Section 7.5**
> **Work**
> **Calculus, 4th Edition,**
> **Larson/Hostetler/Edwards**

Up to this point we have only calculated the work done in moving an object in a straight line path. In the real world, objects frequently move along curved paths. For example, balls that are thrown, batted, or kicked will move along paths that are curved.

EXAMPLE

A Bead Moving Along a Wire

Consider the case of a child's toy consisting of a bead weighing 1 ounce that is constrained to move along a wire that has been bent into the shape of a circle of radius 1 foot. Using the equation for a circle, $x^2 + y^2 = 1$, we wish to calculate the work done by gravity in moving the bead from the point $(0, 1)$ to the point $(1, 0)$ in a clockwise direction. (See Figure 7.11.) We shall assume that the component of the force involved in this work is given by

$$F(x) = \frac{1}{16}x \text{ pounds.}$$

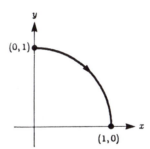

FIGURE 7.11

SOLUTION

The work done in moving a distance ds is given by

$$F(x)\, ds = \frac{1}{16}x\sqrt{1 + \left(\frac{dy}{dx}\right)^2}\, dx.$$

Hence, the total work done is given by

$$\text{Work} = \int_0^1 \frac{1}{16}x\sqrt{1 + \left(\frac{dy}{dx}\right)^2}\, dx.$$

For the circle, $y = \sqrt{1 - x^2}$, and so

$$\frac{dy}{dx} = -\frac{x}{\sqrt{1 - x^2}}.$$

Thus,

$$\sqrt{1 + \left(\frac{dy}{dx}\right)^2} = \sqrt{1 + \frac{x^2}{1 - x^2}} = \sqrt{\frac{1 - x^2 + x^2}{1 - x^2}} = \frac{1}{\sqrt{1 - x^2}}.$$

Hence,

$$\text{Work} = \frac{1}{16}\int_0^1 x(1-x^2)^{-1/2}\,dx.$$

Letting $u = 1 - x^2$, we have $du = -2x\,dx$ which implies that

$$-\frac{1}{2}\,du = x\,dx.$$

Then,

$$\begin{aligned}
\text{Work} &= \frac{1}{16}\int_1^0 u^{-1/2}\left(-\frac{1}{2}\right)\,du \\
&= \frac{1}{32}\int_0^1 u^{-1/2}\,du \\
&= \frac{1}{32}\frac{u^{1/2}}{1/2}\Bigg]_0^1 \\
&= \frac{1}{16}\ \text{ft}\cdot\text{lb}.
\end{aligned}$$

APPLICATION 7.9 EXERCISES

1. Consider an object moving along a parabolic path given by $y = 1 - x^2$ from $(0,\ 1)$ to $(1,\ 0)$. Calculate the work done by the force given by $F(x) = x$ along this path.

2. Consider an object moving along a path given by $y = \frac{1}{3}(x^2 + 2)^{3/2}$ for $0 \le x \le 1$. Calculate the work done by the force given by $F(x) = x$ along this path.

3. An object moves along an elliptical path given by $x = 2\cos t + 1$ and $y = 3\sin t + 2$ for $0 \le t \le \pi/2$. (a) Eliminate t to obtain an equation for this ellipse in terms of x and y. (b) Set up the integral (do not integrate) for the work done by the force given by $F(x) = x$ along this path.

4. A ball that has been thrown moves along a path that is given by $x = 50\sqrt{2}t$ and $y = 50\sqrt{2}t - 32t^2$. (a) Eliminate t to obtain an equation for this curve in terms of x and y. (b) Determine the range for the ball. (c) Set up the integral (do not integrate) for the work done by the force given by $F(x) = x$ along this path.

APPLICATION 7.10: Water Pressure

> See Section **7.6**
> Fluid Pressure and Fluid Force
> Calculus, 4th Edition,
> Larson/Hostetler/Edwards

A column of water whose depth is 33.9 feet has atmospheric pressure at the base of the column of about 14.7 pounds/square inch. Such a column of water corresponds to the maximum height that water can be siphoned. If one were to try to pump water from a well with a pump that "sucks" the water up, the maximum depth that the well could have is therefore 33.9 feet. This type of pump is called a reduced-air-pressure pump. (See Figure 7.12.)

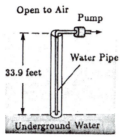

Shallow Well Pump

FIGURE 7.12

EXAMPLE

Reduced-Air-Pressure Pump

Suppose we have a water pump that uses reduced air pressure to draw water out of a well. If the pump can reduce the air pressure to 4.7 pounds/square inch, what is the maximum depth that the well can be?

SOLUTION

We know that atmospheric pressure is about 14.7 pounds/square inch. The pump will permit us to support a column of water with a difference in pressure of 10 pounds/square inch. Simple proportions yield the solution.

$$\frac{h}{33.9} = \frac{10}{14.7}$$
$$h = 23 \text{ feet}$$

Thus, the deepest well from which water can be pumped with this pump is 23 feet.

APPLICATION 7.10 EXERCISES

1. Instead of being siphoned from a well, water can be pumped up by containing it within a chamber and then increasing pressure in the chamber. How deep could a well be if the pump provided a pressure of 100 pounds/square inch above atmospheric pressure?

2. We have considered two pumps: the increased-pressure pump of Exercise 1, and the reduced-pressure pump of the example. If both pumps were used on the same well, find the depth of the well that would be possible.

3. Due to osmotic pressure differences, trees can grow much taller than 33.9 feet and effectively pump water up to the top. The Tall Tree is a coast redwood reportedly more than 367 feet tall. It is the tallest living tree. What pressure at the base would be necessary to support such a water column? (Disregard osmotic pressure.)

4. At five feet above the base, the Tall Tree (see Exercise 3) is a mere 10 feet in diameter. Suppose the xylem layer (the layer that transports water) is 1 inch thick. What is the force due to the fluid on a horizontal cross section of the xylem at this height?

APPLICATION 7.11: Simple Harmonic Motion

> **See Section 7.7**
> **Moments, Centers of Mass, and Centroids**
> **Calculus, 4th Edition,**
> **Larson/Hostetler/Edwards**

In simple harmonic motion, a system vibrates (or oscillates) at a frequency f, or angular frequency ω, where $\omega = 2\pi f$. The angular frequency is given by $\omega = \sqrt{k/m}$, where k is the spring constant for the system. The constant k is a restoring force which is assumed to obey Hooke's Law and m is the mass of the moving (or vibrating) object.

EXAMPLE

A Stiff Spring

One of the most common classroom models of simple harmonic motion is the simple mass-spring system illustrated in Figure 7.13. Typically, the mass of the object is so much greater than that of the spring that the mass of the spring is ignored. We can correct the assumption that the spring has no mass provided we assume that the spring is fairly stiff. This means that the spring motion is uniform, that the entire spring is stretched at the same time, and that the stretching is proportional to the position as measured from the fixed end. A spring that is not stiff would produce propagating waves much like a Slinky. How does the mass of the spring affect the frequency at which the mass-spring system oscillates?

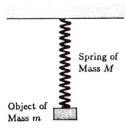

Spring of Mass M

Object of Mass m

FIGURE 7.13

SOLUTION

Our example is for a uniform spring of total mass M and spring constant k. We will approach the problem by using energy considerations. We know the kinetic energy of a moving object with velocity v and mass m is $\frac{1}{2}mv^2$. Then a small amount of our spring mass, Δm_i with speed v_i, has kinetic energy

$$\Delta E_i = \frac{1}{2}\Delta m_i v_i{}^2.$$

With our assumption that the spring is stiff, we have $v_i = V y_i/Y$, where V is the velocity of the object, Y is the coordinate of the object, and y_i is the coordinate of the element of the spring. The mass associated with a small piece of spring is $\Delta m_i = M \Delta y/Y$. This results in

$$E \approx \sum_{i=1}^{n} \Delta E_i$$

$$= \frac{1}{2} \sum_{i=1}^{n} \frac{MV^2 y_i^2}{Y^3} \, \Delta y$$

$$= \frac{1}{2} \int_0^Y \frac{MV^2 y^2}{Y^3} \, dy$$

$$= \frac{1}{2} \left[\frac{MV^2 y^3}{3Y^3} \right]_0^Y$$

$$= \frac{1}{2} \left(\frac{M}{3} \right) V^2.$$

We see then that the kinetic energy of the spring is $\frac{1}{3}$ of the value of the kinetic energy that the spring would have if the entire spring were moving uniformly. From this, we can conclude that the effective mass of the spring is $\frac{1}{3}M$. If we use this in the equation for the angular frequency, we obtain

$$\omega = \frac{1}{2\pi} \sqrt{\frac{k}{m + \frac{1}{3}M}}$$

where m is the mass of the object that is attached to the spring. This relationship gives the frequency in terms of the spring constant k and of the masses of both the spring and the object.

APPLICATION 7.11 EXERCISES

1. Not all springs are uniform. In fact, a nonlinear tapered spring is occasionally used in physics labs in order to maintain a uniform amount of stretching. The idea is that the spring, while hanging, stretches under its own weight. A tapered spring, which is smaller in diameter at the top than at the bottom can be designed in such a way that the vertical spacing in adjacent coils is uniform under the spring's own load and similarly under external loads. Assume that the diameter of the spring increases linearly from r_0 to $2r_0$. Calculate the effective mass of such a spring and the frequency at which the system will oscillate.

2. Consider the spring in Exercise 1. Reverse the direction in which it is used so that the spring's radius decreases linearly from $2r_0$ to r_0. Calculate the effective mass of the spring and the frequency at which the system will oscillate.

3. Two springs are connected together, end to end. A mass of 100 grams is suspended from one end of the pair of springs and the other end of the pair is attached to a support. Both springs are uniform and have the same rest length. The spring on the top has a spring constant of 900 dynes/centimeter and a mass of 20 grams, whereas the spring on the bottom has a spring constant of 625 dynes/centimeter and a mass of 30 grams. Find the frequency of oscillation for this system.

4. Reverse the position of the springs in Exercise 3 and determine the frequency of oscillation.

APPLICATION 7.12: Championship Pole Vault

> **See Section 7.7**
> **Moments, Centers of Mass, and Centroids**
> **Calculus, 4th Edition,**
> **Larson/Hostetler/Edwards**

A quick look at the last ten Olympic Games championship records* for the pole vault (see Table 7.1) show a more-or-less steady increase from a height of 14 feet, $11\frac{1}{8}$ inches to a height of 19 feet, $4\frac{1}{4}$ inches. Olympic records notwithstanding, if we use elementary mechanics to model the pole vault, the results seem to indicate that a vaulter could not clear the heights currently being achieved.

Table 7.1: Olympic Games Championships—Pole Vault

Year	Athlete	Vault Height
1988	Sergei Bubka, Soviet Union	19 feet, $4\frac{1}{4}$ inches
1984	Pierre Quinon, France	18 feet, $10\frac{1}{4}$ inches
1980	Wladyslaw Kozakiewicz, Poland	18 feet, $11\frac{1}{2}$ inches
1976	Tadeusz Slusarski, Poland	18 feet, $\frac{1}{2}$ inches
1972	Wolfgang Nordwig, East Germany	18 feet, $\frac{1}{2}$ inches
1968	Bob Seagren, United States	17 feet, $8\frac{1}{2}$ inches
1964	Fred Hansen, United States	16 feet, $8\frac{3}{4}$ inches
1960	Don Bragg, United States	15 feet, $5\frac{1}{8}$ inches
1956	Robert Richards, United States	14 feet, $11\frac{1}{2}$ inches
1952	Robert Richards, United States	14 feet, $11\frac{1}{8}$ inches

Let's consider a model based on certain assumptions about the sport and the vaulter. We assume that the vaulter is a fast runner who can run 100 yards in 10 seconds, or 30 feet/second. His kinetic energy is therefore $E = \frac{1}{2}mv^2$. If all of that energy (ignoring air resistance) is converted by the pole into raising the pole vaulter, we have

$$\frac{1}{2}mv^2 = mgh$$

$$h = \frac{v^2}{2g}$$

where g is the acceleration due to gravity, which we will take to be 32 feet per second per second. Substitution yields a height of about 14 feet, yet currently people are vaulting 18 or 19 feet!

EXAMPLE

Improving a Pole Vault Model

In our model of the pole vaulter, other factors should be considered so that the model more closely approximates reality.

We have not considered the additional lift that is provided by the vaulter's arms. However, being realistic, we might concede that this could gain perhaps an inch or two, but not the needed 4 or more feet.

---

*Source: *1989 Information Please Almanac*, Houghton Mifflin Company.

A variable not yet considered, however, is the vaulter's position as his center of mass moves across the bar. What effect will this consideration have on our model?

SOLUTION

We are actually concerned with the height change of the center of mass. In the simplest model, we have a person who is like a long stiff rod, of uniform density. If the rod is 6 feet tall, the center of mass would be at 3 feet at the bottom of the jump. If the vaulter was completely horizontal at the top of the jump, he still would not make the 18 feet. If we let the vaulter bend so that he could "slide" over the pole, we then notice that the center of mass will actually pass under the horizontal bar, as shown in Figure 7.14.

When the vaulter is upright, his center of mass is 3 feet above the ground. But when the vaulter is doubled over, his center of mass is $1\frac{1}{2}$ feet from the bottom of the vaulter. The center of mass therefore, moves from a height of 3 feet to a height of $(3 + 14)$ or 17 feet, whereas the top of the vaulter moves to a height of $18\frac{1}{2}$ feet and clears the bar.

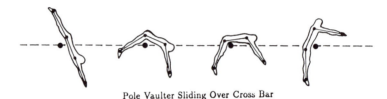

Pole Vaulter Sliding Over Cross Bar

FIGURE 7.14

APPLICATION 7.12 EXERCISES

1. Assume a pole vaulter can double over as he clears the bar. Assume also the vaulter has uniform density. Consider the location of the vaulter's center of mass with reference to the vaulter's lowest point. What is the range of values for the location of the center of mass?

2. Suppose the vaulter is not of uniform density—a more realistic assumption. Let the density of the vaulter be given by the density function $\rho(h) = \rho_0(1 + 0.04h)$, where h is the distance from the vaulter's feet toward his head. Find the range of values for the center of mass for the vaulter when he is standing straight up, doubled over, or upside down.

3. Assume that the pole vaulter with the density described in Exercise 2 can bend like a piece of rubber. The center of mass of the vaulter describes a trajectory during the vault. What is the peak height of that trajectory?

4. Suppose the pole vaulter in the example runs 1 ft/sec faster. What increase in height will he attain?

8

INTEGRATION TECHNIQUES, L'HÔPITAL'S RULE, AND IMPROPER INTEGRALS

APPLICATION 8.1: Spring Problem

> See Section 8.1
> Basic Integration Formulas
> Calculus, 4th Edition,
> Larson/Hostetler/Edwards

The study of oscillations is important in understanding earthquakes and in designing earthquake resistant structures. We consider the problem of a spring which has been set in motion and thus begins to oscillate.

EXAMPLE

A Spring Problem

A mass m of 1 kilogram is attached to a spring with spring constant $k = 200$ Newtons/meter. (See Figure 8.1.) We assume that there is no air resistance and that the spring has come to an equilibrium position at $x = 0$. The spring is then pulled down to an initial position $x_0 = \sqrt{2}/2$ meters and released with an initial velocity $v_0 = 0$.

(a) Find an equation relating x and v.

(b) Find the position at any time t.

Problems such as this are usually solved using more advanced techniques such as differential equations.* We shall solve this problem using only integration.

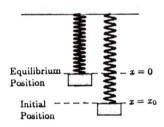

Equilibrium Position — $x = 0$

Initial Position — $x = x_0$

FIGURE 8.1

SOLUTION

We assume Hooke's Law,

$$F = ma = m\frac{dv}{dt} = -kx$$

which implies that $dv/dt = -200x$.

(a) We begin by rewriting the acceleration as

$$\frac{dv}{dt} = \frac{dv}{dx}\frac{dx}{dt} = v\frac{dv}{dx}$$

which implies that

$$v\frac{dv}{dx} = -200x.$$

---

*See *Calculus,* Larson/Hostetler/Edwards, 4th Edition, D. C. Heath and Company, 1990, Chapter 17.

Separating variables, we obtain

$$\int v\, dv = \int -200x\, dx$$

$$\frac{1}{2}v^2 = -100x^2 + C.$$

Using the initial conditions $v = 0$ when $x = \sqrt{2}/2$, we get $C = 50$. Thus,

$$\frac{1}{2}v^2 + 100x^2 = 50$$

which is in accord with the law of conservation of energy (the sum of the kinetic energy and the potential energy is constant).

(b) We wish to solve for the position x as a function of time. We first solve the energy equation for v:

$$v = \pm\sqrt{100 - 200x^2}.$$

Thus,

$$v = \frac{dx}{dt} = \pm\sqrt{100 - 200x^2}$$

$$\frac{dx}{\sqrt{100 - 200x^2}} = dt$$

$$\int \frac{dx}{\sqrt{100 - 200x^2}} = t$$

$$\frac{1}{10\sqrt{2}} \int \frac{du}{\sqrt{100 - u^2}} = t$$

$$\frac{1}{10\sqrt{2}} \sin^{-1}\left(\frac{u}{10}\right) + C_1 = t$$

where $u = \sqrt{200}\, x = 10\sqrt{2}\, x$. Applying the initial condition $x = \sqrt{2}/2$ when $t = 0$, we obtain $C_1 = -\pi/20\sqrt{2}$. Therefore,

$$\frac{1}{10\sqrt{2}} \sin^{-1}[\sqrt{2}\, x] = t + \frac{\pi}{20\sqrt{2}}$$

and we can write

$$x = \frac{1}{\sqrt{2}} \sin\left(10\sqrt{2}\, t + \frac{\pi}{2}\right).$$

APPLICATION 8.1 EXERCISES

1. Solve the example except that the initial conditions are $v_0 = 0$ and $x_0 = 1$ meter when $t = 0$.

2. Solve the example except that the initial conditions are $v_0 = 10$ meters/second and $x_0 = 1$ meter when $t = 0$.

3. Solve the example with initial conditions $v_0 = 0$ and $x_0 = 0$ when $t = 0$ and interpret the results.

4. Solve the example with mass m, spring constant k, and initial conditions $x = x_0$ and $v = v_0$.

5. For small angles θ, the motion of a pendulum of length L can be described by the equation

$$\frac{d^2\theta}{dt^2} + \frac{g}{L}\theta = 0$$

where g is the acceleration due to gravity. (See figure.) Solve the equation for θ subject to the initial conditions $t = 0$, $\theta = \pi/12$, and $\theta' = 0$ where $L = 2$ feet and $g = 32$ ft/sec^2.

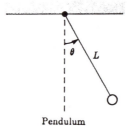

Pendulum

FIGURE FOR EXERCISE 5

6. A torsional pendulum consists of a rigid body suspended by a wire attached to a support. (See figure.) When the body is twisted through an angle θ, the twisted wire exerts a restoring torque. The balance wheel of a clock behaves in this manner. The motion can be described by the equation

$$\frac{d^2\theta}{dt^2} = -\frac{K}{I}\theta$$

where K is the torsion constant and I is the moment of inertia of the body. Solve the equation for a clock with $K/I = 4\pi^2$, subject to the initial conditions $t = 0$, $\theta = \pi/6$, and $\theta' = 0$.

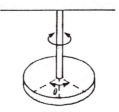

Torsional Pendulum

FIGURE FOR EXERCISE 6

APPLICATION 8.2: The Egyptian Water Clock

> See Section 8.1
> **Basic Integration Formulas**
> **Calculus, 4th Edition,**
> **Larson/Hostetler/Edwards**

An Egyptian water clock is a large container with a hole in the bottom. It is filled daily with 24 hours worth of water which slowly leaks out. Markings on the side of the container are spaced to indicate the passage of each hour.

Torricelli (1609–1647) stated the principle on which the water clock is based. Torricelli's Law states that the rate at which water, or other low-viscosity liquid, flows out of an open container is the same as the rate the water would have acquired if it had been in free fall and had fallen a height equivalent to that of the fluid above the opening of the container. It can easily be shown that Torricelli's Law implies that the volume of liquid within a container with a hole in it decreases proportionally to the distance the hole is beneath the surface of the liquid.

The change in volume can be thought of as a small cylinder of liquid coming out of the container. (See Figure 8.2.) In other words,

$$dV = -A\,dh$$

$$\frac{dV}{dt} = -A\frac{dh}{dt} = -Av$$

where A is the area of a cross section of the cylinder, and v is the speed of the flowing fluid. For free fall, we know that $v^2 = 2gh$ so we can conclude that

$$\frac{dV}{dt} = -A\sqrt{2gh}$$

which is the desired result.

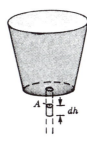

FIGURE 8.2

EXAMPLE

Marking the Hours

Consider an Egyptian water clock in the shape of a right circular cylinder. Use Torricelli's Law to determine where the marks on the container should be placed to indicate each passing hour.

SOLUTION

Torricelli's Law states that

$$\frac{dV}{dt} = -k\sqrt{h}$$

where k is a constant, V is the volume, and h is the current depth of the liquid, as shown in Figure 8.3.

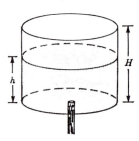

FIGURE 8.3

For our cylinder, $V = \pi r^2 h$ and $dV/dt = \pi r^2 (dh/dt)$. Equating the two expressions for dV/dt, we obtain

$$\pi r^2 \frac{dh}{dt} = -k\sqrt{h}$$

or

$$\frac{dh}{dt} = -\alpha \sqrt{h}$$

where $\alpha = k/\pi r^2 =$ constant. We can rewrite this equation as

$$\frac{dh}{\sqrt{h}} = -\alpha \, dt$$

and then integrate to obtain

$$2\sqrt{h} = -\alpha t + C.$$

We still need to evaluate the two constants, α and C. Two conditions which enable us to complete this problem are that at $t = 0$, $h = H$, and at $t = 24$, $h = 0$. At $t = 0$, we get

$$2\sqrt{H} = C$$

and at $t = 24$ we get

$$0 = -24\alpha + C$$
$$= -24\alpha + 2\sqrt{H}.$$

We can then rewrite our result as

$$2\sqrt{h} = -\frac{1}{12}\sqrt{H}\,t + 2\sqrt{H}$$
$$2\sqrt{h} = \left(2 - \frac{t}{12}\right)\sqrt{H}$$
$$\sqrt{h} = \left(1 - \frac{t}{24}\right)\sqrt{H}$$
$$h = \left(1 - \frac{t}{24}\right)^2 H.$$

The hour marks on the cylindrical container should then be placed at each h where t is replaced by the integers 1 through 23. Of course, 0 corresponds to full and the beginning of the day, and 24 corresponds to empty and the end of the day.

APPLICATION 8.2 EXERCISES

1. Consider the Egyptian water clock of the example. If the container were a cone with vertex pointed down, where should the marks be placed?

2. A container which is a paraboloid of revolution is to be used as an Egyptian water clock. Find the location of the hour marks. (The revolved parabola is $y = x^2$, $0 \leq x \leq 1$.) See figure.

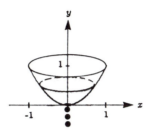

FIGURE FOR EXERCISE 2

3. Determine a shape for a container such that the height of the fluid in a vessel falls at a constant rate. Assume Torricelli's Law: $dV/dt = -k\sqrt{h}$.

4. For the container in Exercise 3, find the location of the hour markings.

APPLICATION 8.3: Relativity

> See Section 8.1
> **Basic Integration Formulas**
> **Calculus, 4th Edition,**
> **Larson/Hostetler/Edwards**

When an object travels near the speed of light, we are in the realm of relativity theory. Newton's Second Law of Motion, given by $F = ma$, applies only to the case of constant mass, but relativity theory tells us that mass near these high velocities increases significantly. For problems involving relativity, we use the more general formula given by

$$F = \frac{d}{dt}[mv].$$

The mass m is related to the rest mass m_0 by

$$m = \frac{m_0}{\sqrt{1 - (v^2/c^2)}}$$

where c is the speed of light and v is the velocity of the mass.

EXAMPLE

Superfast Rocket

Assume we have a rocket whose motion we are analyzing by relativity theory. The rocket has a rest mass of 1000 kilograms and an unlimited supply of fuel.

(a) If the force applied to the rocket is constant and the rocket starts from rest, determine the velocity and position of the rocket at any time t.

(b) Determine the $\lim_{t \to \infty} v(t)$.

SOLUTION

(a) We begin by determining the velocity of the rocket. Let $F = k$. Then,

$$k = \frac{d}{dt}\left[\frac{m_0 v}{\sqrt{1 - (v^2/c^2)}}\right] = \frac{d}{dt}\left[\frac{1000v}{\sqrt{1 - (v^2/c^2)}}\right].$$

Integration yields

$$kt + C_1 = \frac{1000v}{\sqrt{1 - (v^2/c^2)}}.$$

Using the initial condition $v = 0$ when $t = 0$, we obtain $C_1 = 0$. Thus,

$$kt = \frac{1000v}{\sqrt{1 - (v^2/c^2)}}$$

$$k^2 t^2 = \frac{1000^2 v^2}{1 - (v^2/c^2)} = \frac{c^2 1000^2 v^2}{c^2 - v^2}$$

$$c^2 k^2 t^2 - v^2 k^2 t^2 = c^2 1000^2 v^2$$

$$v^2 = \frac{k^2 c^2 t^2}{c^2 1000^2 + k^2 t^2}$$

$$v = \frac{kct}{\sqrt{c^2 1000^2 + k^2 t^2}}.$$

This is the velocity of the rocket as a function of time. We now determine the position of the rocket.

$$v = \frac{ds}{dt} = \frac{kct}{\sqrt{c^2 1000^2 + k^2 t^2}}$$

$$s = \int \frac{kct\, dt}{\sqrt{c^2 1000^2 + k^2 t^2}}$$

Let $u = c^2 1000^2 + k^2 t^2$ and $du = 2k^2 t\, dt$. Then,

$$s = \int kcu^{-1/2} \frac{du}{2k^2} = \frac{c}{2k} \frac{u^{1/2}}{1/2} + C_2 = \frac{c}{k} \sqrt{c^2 1000^2 + k^2 t^2} + C_2.$$

Using the initial condition $s = 0$ when $t = 0$, we obtain

$$C_2 = -\frac{c^2}{k} 1000.$$

Hence,

$$s = \frac{c}{k} \sqrt{c^2 1000^2 + k^2 t^2} - \frac{c^2 1000}{k}.$$

This is the position of the rocket as a function of time.

(b) We now evaluate

$$\lim_{t \to \infty} v(t) = \lim_{t \to \infty} \frac{kct}{\sqrt{c^2 1000^2 + k^2 t^2}}.$$

Dividing the numerator and the denominator by t, we obtain

$$\lim_{t \to \infty} v(t) = \lim_{t \to \infty} \frac{kc}{\sqrt{[(c^2 1000^2)/t^2] + k^2}} = \frac{kc}{\sqrt{k^2}} = c.$$

Thus, the speed of light represents the limiting velocity for this problem Certainly, without relativity, an object acted upon by a constant force (no friction) would approach an infinite velocity.

APPLICATION 8.3 EXERCISES

1. Try to evaluate

$$\lim_{t \to \infty} \frac{kct}{\sqrt{c^2 1000^2 + k^2 t^2}}$$

using L'Hôpital's Rule. Is L'Hôpital's Rule helpful for this problem?

2. Reconsider the example for the case in which force is decreasing and is given by $F(t) = k/(t+1)$. Determine the limit $\lim_{t \to \infty} v(t)$.

3. Reconsider the example for the case in which the force is given by $F(t) = k(t+1)^2$. Determine the limit $\lim_{t \to \infty} v(t)$.

4. Reconsider the example for the case in which the force is given by $F(t) = k/(t+1)^{1/2}$. Determine the limit $\lim_{t \to \infty} v(t)$.

APPLICATION 8.4: Quantum Mechanics

> See Section 8.2
> **Integrating by Parts**
> Calculus, 4th Edition,
> Larson/Hostetler/Edwards

The results that are obtained in solving problems in quantum mechanics are obtained by solving Schrödinger's Equation, which is the quantum mechanical equivalent of Newton's Law. A notable difference is that the result from Newton's Law is usually a trajectory whereas solutions to Schrödinger's Equation represent a wave function. This is the "probability amplitude." The wave function alone is not meaningful. To find a physical value, we need to treat the wave function in ways similar to what might be done in statistics.

EXAMPLE

Particle in a Box

In the study of quantum mechanics, one problem which can be solved exactly is the particle in a box. This problem concerns a very small object contained within a box and not interacting with anything else. The model states that a particle will move about the box but cannot tunnel through the walls and escape. What are the expected values of the momentum and kinetic energy of our particle in a box?

SOLUTION

The wave function for a one-dimensional particle in a box of length L is

$$\psi(x) = \sqrt{\frac{2}{L}} \sin \frac{n\pi x}{L}, \qquad 0 < x < L.$$

You will first notice that the particle is indeed in the box since

$$\int_0^L [\psi(x)]^2 \, dx = \frac{2}{L} \int_0^L \sin^2 \frac{n\pi x}{L} \, dx = 1.$$

In other words, the probability of finding the particle in the box is 1. Where are we most likely to find the particle within the box? Here we need only calculate the value of the integral

$$\int_0^L x[\psi(x)]^2 \, dx = \frac{2}{L} \int_0^L x \sin^2 \frac{n\pi x}{L} \, dx.$$

One would expect the average momentum to be zero. This calculation can also be done quite easily by use of quantum mechanics. Momentum is obtained by using the "momentum operator" which is given by $p = -i\hbar(d/dx)$ where $\hbar$ (read as h-bar) is Planck's constant divided by 2π and i is the usual $\sqrt{-1}$. It is fortunate that the resulting integral is zero or else we would have obtained an imaginary result.

Next we can calculate the expected kinetic energy of a particle in a box. Normally we would write kinetic energy as $\frac{1}{2}mv^2$ but, in quantum mechanics, momentum is the variable of choice, so the kinetic energy is $p^2/2m$. We already know that $p = -i\hbar(d/dx)$ so the kinetic energy operator is $(-\hbar^2/2m)(d^2/dx^2)$. Therefore, the kinetic energy is

$$\text{kinetic energy} = \frac{2}{L} \int_0^L \sin \frac{n\pi x}{L} \left(-\frac{\hbar^2}{2m} \right) \frac{d^2}{dx^2} \left[\sin \frac{n\pi x}{L} \right] dx$$

$$= \frac{2}{L} \frac{(n\pi\hbar)^2}{2mL^2} \int_0^L \sin \frac{n\pi x}{L} \sin \frac{n\pi x}{L} dx$$

$$= \frac{(n\pi\hbar)^2}{2mL^2}.$$

APPLICATION 8.4 EXERCISES

1. Complete the integral for the average position for a particle in a box.

2. Show that the integral for the average momentum is indeed zero.

3. Find the most probable position for a particle trapped in a harmonic oscillator. The wave function for the lowest energy level of a harmonic oscillator is

$$\psi(x) = \left(\frac{\alpha^{1/2}}{\pi^{1/4}} \right) e^{-\alpha^2 x^2/2}, \quad -\infty < x < \infty$$

where $\alpha^4 = mk/\hbar^2$ and k is the spring constant.

4. Find the expected kinetic energy of the harmonic oscillator. (See Exercise 3.)

APPLICATION 8.5: *RMS* **Values**

> **See Section 8.3**
> **Trigonometric Integrals**
> **Calculus, 4th Edition,**
> **Larson/Hostetler/Edwards**

Power companies in the United States provide residential users with 110 volts, 60 Hertz electricity. But what does this really mean?

We use alternating current (AC), which is continuously changing. The measurement 60 Hertz means that the electricity can be represented by a sine function with a period of 1/60 of a second. The 110 volts is what is called an *rms* (root mean squared) value. To find a root mean square value, one calculates the square root of the average value of the square of a quantity.

EXAMPLE

Wall Outlet Voltage

In the United States, the voltage of the electric wall socket can be described by

$$v(t) = v_0 \sin(120\pi t + \phi)$$

where ϕ is a phase angle that depends on t_0, the time at which we start our clock measuring the voltage. Use the root mean square concept to obtain the value for v_0.

SOLUTION

We want first the average of v^2 for one cycle (the mean squared value) and then we will take the square root. This value, the *rms* value, we know is 110. We start with the mean squared value,

$$\frac{1}{T}\int_0^T v^2(t)\, dt.$$

Since one cycle is 1/60 of a second, and voltage is $v_0 \sin(120\pi t + \phi)$, our integral becomes

$$\frac{1}{T}\int_0^T v^2(t)\, dt = \frac{1}{1/60}\int_0^{1/60} v_0{}^2 \sin^2(120\pi t + \phi)\, dt$$

$$= 30v_0{}^2 \int_0^{1/60} \left[1 - \cos(240\pi t + 2\phi)\right] dt$$

$$= 30v_0{}^2 \left[t - \frac{\sin(240\pi t + 2\phi)}{240\pi}\right]_0^{1/60}$$

$$= 30v_0{}^2 \left[\frac{1}{60} - \frac{\sin(4\pi + 2\phi)}{240\pi} + \frac{\sin 2\phi}{240\pi}\right] = \frac{v_0{}^2}{2}.$$

Now we take the square root and equate the result to the known value, 110.

$$\sqrt{\frac{v_0{}^2}{2}} = \frac{v_0}{\sqrt{2}} = 110$$

This implies that, $v_0 = 110\sqrt{2}$, which is the value we sought. The voltage function can now be written as

$$v(t) = 110\sqrt{2}\sin(120\pi t + \phi).$$

APPLICATION 8.5 EXERCISES

1. Suppose electricity is processed by an electronic component called a rectifier, which only lets $\frac{1}{2}$ of the alternating current flow past. The resulting voltage for one cycle is

$$v(t) = \begin{cases} v_0 \sin 120\pi t, & 0 < t < \frac{1}{120} \text{ seconds} \\ 0, & \frac{1}{120} < t < \frac{1}{60} \text{ seconds.} \end{cases}$$

   Find the *rms* value for $v(t)$.

2. Find the *rms* value for a square wave given by

$$f(t) = \begin{cases} 1, & 0 < t < \frac{T}{2} \\ 0, & \frac{T}{2} < t < T. \end{cases}$$

3. Find the *rms* value for a sawtooth wave given by $f(t) = t, 0 < t < T$, and $f(t + T) = f(t)$.

4. Besides being used in the study of electrical waves, root mean square calculations are used in other topics. For example, we can determine the *rms* velocity of a gas molecule. In an ideal non-interacting gas, the velocities of the gas molecules are distributed according to the rule derived by Maxwell

$$N(v) = 4\pi N_0 \left(\frac{m}{2\pi kT} \right)^{3/2} v^2 e^{-v^2/2kT}, \qquad 0 < v < \infty.$$

   Calculate the *rms* value of the velocity of a typical gas molecule.

APPLICATION 8.6: Pursuit Problem

> See Section 8.4
> Trigonometric Substitution
> Calculus, 4th Edition,
> Larson/Hostetler/Edwards

EXAMPLE

The Dog and Rabbit Problem

A dog spots a rabbit running in a straight line in a field. Fortunately for the rabbit, the distance between the dog and the rabbit remains constant. Assuming that the dog always moves toward the rabbit and that the distance between them is 10 feet, find an equation to describe the path of the dog.

SOLUTION

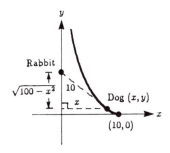

FIGURE 8.4

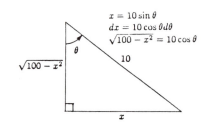

FIGURE 8.5

See Figure 8.4. Let $(0, 0)$ and $(10, 0)$ represent the initial positions of the rabbit and the dog, respectively, and assume that the rabbit moves along the y-axis. Let (x, y) represent the position of the dog. We wish to establish a differential equation involving dy/dx. The slope of the tangent line to the dog's path is

$$\frac{dy}{dx} = -\frac{\sqrt{100 - x^2}}{x}.$$

Using trigonometric substitution (as indicated in Figure 8.5), we solve for y as follows.

$$y = -\int \frac{\sqrt{100 - x^2}}{x} \, dx$$

$$= -\int \frac{10 \cos \theta}{10 \sin \theta} 10 \cos \theta \, d\theta$$

$$= -10 \int \frac{1 - \sin^2 \theta}{\sin \theta} \, d\theta$$

$$= -10 \int \left(\csc \theta - \sin \theta \right) d\theta$$

$$= -10[-\ln |\csc \theta + \cot \theta| + \cos \theta] + C$$

$$= 10 \left[\ln \left| \frac{10}{x} + \frac{\sqrt{100 - x^2}}{x} \right| - \frac{\sqrt{100 - x^2}}{10} \right] + C.$$

Next, we use the initial condition $y = 0$ when $x = 10$. So,

$$0 = 10 \ln 1 + C$$
$$0 = C.$$

Hence, the equation for the dog's path is

$$y = 10 \left[\ln \left| \frac{10}{x} + \frac{\sqrt{100 - x^2}}{x} \right| - \frac{\sqrt{100 - x^2}}{10} \right]$$

$$= 10 \ln \left| \frac{10 + \sqrt{100 - x^2}}{x} \right| - \sqrt{100 - x^2}.$$

APPLICATION 8.6 EXERCISES

1. Solve the example as given except assume that the dog starts at the point $(0, 10)$ and that the rabbit moves along the x-axis.

2. Solve the example as given except assume that the dog starts at the point $(9, 0)$, that the dog always moves toward the rabbit, and that the speed of the dog is twice the speed of the rabbit.* Find an equation to describe the path of the dog.

3. Solve Exercise 2 except assume that the dog starts at the point $(4, 0)$ and that the rabbit moves along the y-axis.

---

*If the speed of the dog is twice the speed of the rabbit, the dog will catch the rabbit. See *Calculus*, Larson/Hostetler/Edwards, 4th Edition, D. C. Heath and Company, 1990, page 436.

APPLICATION 8.7: Falling Object

> **See Section 8.4**
> **Trigonometric Substitution**
> **Calculus, 4th Edition,**
> **Larson/Hostetler/Edwards**

A space station moving around the earth may send a "package" down to the surface either by crew action or by a mechanical drop. Although it may seem that only high-speed computer networks could calculate when such a package hits the ground, in actuality, pencil-and-paper calculations using some calculus are all that is needed.

For example, the acceleration due to gravity will determine a differential equation that relates the distance x the package is above the earth, and its velocity. Solving this differential equation yields another equation relating position and time. By patiently manipulating the sequence of equations and making use of initial conditions, we can determine the time it takes for the dropped object to hit the earth. We will ignore air resistance.

EXAMPLE

Object Dropped to Earth

An instrument package is dropped from a height of 4000 miles above the surface of the earth. It is acted upon only by the force due to gravity. If the radius of the earth is 4000 miles, determine the time it takes for the object to hit the earth.

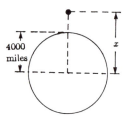

FIGURE 8.6

SOLUTION

Let x represent the distance to the center of the earth. The force due to gravity is given by

$$F = -\frac{k}{x^2} = ma = m\frac{dv}{dt}$$

which implies that

$$\frac{dv}{dt} = -\frac{C}{x^2}.$$

Using $a = -g$ when $x = 4000$, we have $C = (4000)^2 g$. Now, since

$$\frac{dv}{dt} = \frac{dv}{dx}\frac{dx}{dt} = v\frac{dv}{dx}$$

we obtain

$$v\frac{dv}{dx} = -\frac{C}{x^2}$$

subject to the initial conditions $t = 0$, $v = 0$, and $x = 8000$. We next separate variables to obtain

$$v \, dv = -C \frac{dx}{x^2}.$$

Integration yields

$$\frac{v^2}{2} = \frac{C}{x} + C_1.$$

Applying the initial conditions gives

$$0 = \frac{C}{8000} + C_1$$

which implies that $C_1 = -C/8000$. Therefore,

$$\frac{v^2}{2} = \frac{C}{x} - \frac{C}{8000}$$

$$v = -\sqrt{\frac{2C}{x} - \frac{2C}{8000}}$$

$$\frac{dx}{dt} = -\sqrt{\frac{8000 - x}{8000x}} \sqrt{2C}.$$

We separate variables once more to obtain

$$\int \sqrt{\frac{8000x}{8000 - x}} \, dx = -\int \sqrt{2C} \, dt.$$

Then, using the trigonometric substitution $\sqrt{x} = \sqrt{8000} \sin \theta$, we obtain

$$(8000)^{3/2} \left[\sin^{-1} \frac{\sqrt{x}}{\sqrt{8000}} - \frac{\sqrt{x}\sqrt{8000 - x}}{8000} \right] + C_2 = -\sqrt{2C}\, t.$$

From the initial conditions we see that

$$C_2 = -(8000)^{3/2} \frac{\pi}{2}.$$

Thus,

$$t = \frac{(8000)^{3/2}}{\sqrt{2C}} \left[\frac{\pi}{2} - \sin^{-1} \frac{\sqrt{x}}{\sqrt{8000}} + \frac{\sqrt{x}\sqrt{8000 - x}}{8000} \right].$$

Finally, we set $x = 4000$ and obtain

$$t = \frac{(8000)^{3/2}}{\sqrt{2C}} \left[\frac{\pi}{2} - \sin^{-1} \sqrt{\frac{1}{2}} + \frac{1}{2} \right]$$

$$= \frac{(8000)^{3/2}}{\sqrt{2C}} \left[\frac{\pi}{4} + \frac{1}{2} \right]$$

$$= \frac{(8000)^{3/2}}{\sqrt{2g}4000} \left[\frac{\pi}{4} + \frac{1}{2} \right].$$

The units depend on the units of time chosen for g. (Time will be measured in hours if the value of g is taken to be $g = 78,545$ miles/hour^2.)

APPLICATION 8.7 EXERCISES

1. Solve the example as given except assume that the object is dropped from a height of 1000 miles above the surface of the earth.

2. Solve the integral

$$\int \sqrt{\frac{8000x}{8000 - x}}\, dx$$

   using the substitution $\sqrt{x} = \sqrt{8000}\cos\theta$.

3. Solve the example as given except assume that the object is dropped from a height of 1000 miles above the surface of the moon. Assume that the radius of the moon is 1080 miles and that the acceleration due to gravity at the surface of the moon is $0.17g$.

4. Solve the example as given except assume that the object is dropped from a height of 1000 miles above the surface of the planet Mars. Assume that the radius of Mars is 2100 miles and that the acceleration due to gravity at the surface of Mars is $0.38g$.

APPLICATION 8.8: Mellin Transform

> See Section 8.8
> **Improper Integrals**
> **Calculus, 4th Edition,**
> **Larson/Hostetler/Edwards**

An important branch of mathematics involves integral transforms. The basic idea is to take a difficult problem and convert (transform) it into a problem that can be solved more easily. The most common transforms are the Laplace transform, the Fourier transform, the Hankel transform, and the Mellin transform. It is important to have a variety of transforms because some problems can be solved more simply using one type of transform than another. In this example, we consider the Mellin transform which can be used to solve certain partial differential equations such as Laplace's Equation in the wedge-shaped region $\{(r, \theta) : r > 0, \ -\alpha < \theta < \alpha\}$. The Mellin transform is defined by

$$M(f)(s) = \int_0^\infty f(x) x^{s-1} \, dx.$$

We shall assume that s is a real number although in practice s is permitted to be a complex number.

EXAMPLE

Mellin Transform

Use the Mellin transform to solve the equation

$$x f'(x) + f(x) = g(x) = \begin{cases} 0, & \text{if } 0 \le x < 1 \\ 1, & \text{if } 1 \le x \end{cases}$$

with the condition that $\lim_{x \to 0} x^s f(x) = 0$ for $s < 0$.

SOLUTION

We apply the Mellin transform to the equation.

$$\begin{aligned}
M(g)(s) &= \int_0^\infty g(x) x^{s-1} \, dx \\
&= \int_0^1 (0)(x^{s-1}) \, dx + \int_1^\infty (1)(x^{s-1}) \, dx \\
&= \lim_{b \to \infty} \int_1^b x^{s-1} \, dx \\
&= \lim_{b \to \infty} \frac{x^s}{s} \Big]_1^b \\
&= \lim_{b \to \infty} \left(\frac{b^s}{s} - \frac{1}{s} \right) \\
&= -\frac{1}{s} \text{ if } s < 0.
\end{aligned}$$

So, $M(g)(s) = -1/s$. Applying integration by parts, we see that

$$M[x f'(x)](s) = -s M(f)(x).$$

(See Exercise 3.) We can now apply the Mellin transform to the equation to obtain

$$-s\mathcal{M}(f) + \mathcal{M}(f) = -\frac{1}{s}$$

which implies that

$$\mathcal{M}(f) = \frac{1}{s(s-1)} = \frac{1}{s-1} - \frac{1}{s}.$$

Thus,

$$f(x) = \begin{cases} 0, & \text{if } 0 \le x < 1 \\ 1 - \frac{1}{x}, & \text{if } 1 \le x. \end{cases}$$

(See Exercise 2.)

APPLICATION 8.8 EXERCISES

1. Let

$$f(x) = \begin{cases} 0, & 0 \le x \le 1 \\ 1, & 1 < x < 2 \\ 2, & 2 \le x. \end{cases}$$

Evaluate $\mathcal{M}(f)$ and determine the values of s for which this integral is infinite.

2. Solve the problem in Exercise 1 for the function

$$f(x) = \begin{cases} 0, & 0 < x < 1 \\ \frac{1}{x}, & 1 \le x. \end{cases}$$

3. Show that $\mathcal{M}[xf'(x)](s) = -s\mathcal{M}(f)(s)$ if $\lim_{x \to 0} x^s f(x) = 0$ for $s < 0$.

4. Show that $\mathcal{M}[f(ax)](s) = a^{-s}\mathcal{M}(f)(s)$.

5. Show that $\mathcal{M}[f(x^a)](s) = a^{-1}\mathcal{M}(f)(s/a)$.

6. Show that $\dfrac{d}{ds}[\mathcal{M}(f)](s) = \mathcal{M}[\ln x f(x)](s)$.

9

INFINITE SERIES

APPLICATION 9.1: Sequences

> **See Section 9.1**
> **Sequences**
> **Calculus, 4th Edition,**
> **Larson/Hostetler/Edwards**

By a configuration of certain types of mirrors, we can produce sequences of images. Some flat-mirror configurations result in a finite sequence of images; some result in an infinite sequence of images. The physics involved is in the realm of geometric optics. The reflected image appears to be directly behind the mirror or the extension of the mirror from where the object is located. An image in one mirror can then behave as an object for another mirror. Mirrors do not reflect all the light incident upon them. As a reasonable estimate, about 90% of the incident light is reflected.

EXAMPLE

Two Flat Mirrors

We consider an arrangement of two flat mirrors at right angles to each other, as shown in Figure 9.1. What is the sequence of images created in the mirrors?

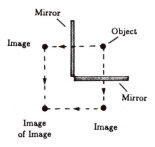

FIGURE 9.1

SOLUTION

We notice here that the sequence of images is finite. In fact, there are only three images, of which one is the image of an image.

APPLICATION 9.1 EXERCISES

1. Two mirrors are mounted on walls parallel to each other 10 meters apart (as in a typical barber shop). You are 4 meters from one mirror and facing it. Write down the locations of the first four images you see of yourself and identify whether each is your face or the back of your head.

2. Suppose each reflection for the arrangement of mirrors in Exercise 1 is only 90% of the incident light. How many reflections will occur before the light intensity is reduced to 1% of the initial value?

3. Let the angle between two flat mirrors be 60°. Find the sequence of images. Is the sequence finite?

4. Let the angle between the two mirrors be θ. Are there any restrictions on θ which will result in a finite number of images?

5. In music, two types of sequences are used to create the 12-note scale. An octave on a piano covers the range of allowed frequencies from f_0 to $2f_0$. The frequency of each note on the scale is $\sqrt[12]{2}$ times the frequency of the prior note. Therefore, $2f_0 = \left(\sqrt[12]{2}\right)^{12} f_0$. On other instruments, the notes are in the ratio of the integers 2, 3, and 5, or powers of these integers. Determine the ratios needed to create a scale that will be close to the piano scale. [Hint: The messiest fraction is $\frac{36}{25}$. All but one of the rest of the ratios have single-digit denominators.]

APPLICATION 9.2: Relativistic Velocity

| See Section 9.1 |
| Sequences |
| Calculus, 4th Edition, |
| Larson/Hostetler/Edwards |

Suppose an object is observed to be traveling at a constant velocity u with respect to an observer who is on Reference Frame 1. Reference Frame 1 is itself moving at a constant velocity v with respect to Reference Frame 2. (See Figure 9.2.) As these two velocities approach the speed of light c, where $|u/c| < 1$ and $|v/c| < 1$, we enter the realm of special relativity. (Note that we assume that the object and both reference frames have parallel, straight-line paths.)

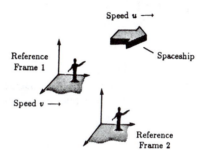

FIGURE 9.2

We can transform the velocity from one reference frame to another. For instance, in this problem the apparent speed w of the object with respect to Reference Frame 2 is given by

$$w = \left(\frac{\dfrac{u}{c} + \dfrac{v}{c}}{1 + \dfrac{uv}{c^2}} \right) c.$$

Note the similarity of this relationship to the addition rule for hyperbolic tangents.

EXAMPLE

Calculating Speed in a Moving Reference Frame

Find the apparent speed of an object that has a relative speed of $\frac{1}{2}c$ in a moving reference frame that is moving at $\frac{1}{2}c$ with respect to a second reference frame.

SOLUTION

Since $u = v = c/2$, we have

$$w = \left(\frac{\dfrac{u}{c} + \dfrac{v}{c}}{1 + \dfrac{uv}{c^2}} \right) c$$

$$= \left(\frac{\dfrac{1}{2} + \dfrac{1}{2}}{1 + \dfrac{1}{4}} \right) c$$

$$= \frac{4}{5} c.$$

APPLICATION 9.2 EXERCISES

1. Find the apparent speed of an object that has a relative speed of 0.6c in a moving reference frame that is moving at 0.4c with respect to a second reference frame.

2. Consider three reference frames. An object is moving at 0.5c in the first reference frame. The first reference is then moving at 0.5c compared to the second reference frame. Finally, the second reference frame is moving at 0.5c compared to the third frame. How fast does the object appear to be moving in the third reference frame?

3. Repeat the transformations of Exercise 2 except assume that all three speeds are 0.9c.

4. Repeat Exercise 2, except assume that there are n reference frames. What is the apparent speed in the nth reference frame?

5. Show that if $|u/c| < 1$ and $|v/c| < 1$, then

$$\left| \frac{\dfrac{u}{c} + \dfrac{v}{c}}{1 + \dfrac{uv}{c^2}} \right| < 1.$$

6. Show that a compound transformation such as that illustrated in Exercise 4 will result in a speed less than c no matter how many transformations occur and will be independent of the relative velocities.

APPLICATION 9.3: The Bohr Model

> See Section 9.1
> Sequences
> Calculus, 4th Edition,
> Larson/Hostetler/Edwards

The founder of our modern theory of atomic structure was Niels Bohr (1885–1962), a Danish theoretical physicist. Bohr applied a primitive quantum theory to the hydrogen atom. His model, called the Bohr model, conceived of the atom as having electrons that travel around a nucleus in orbits affected by quanta of energy.

If a quantum of energy (in the form of light or heat or some other form) excites an electron, it may jump to a higher level (orbit) around the nucleus. As the electron drops back to its previous level, the quantum of energy is released. The energy released, in the form of photons, corresponds to wavelengths of a certain frequency. This yields the so-called atomic spectrum. In this way, the Bohr model leads to the description of the atomic spectra.

As an electron in an atom drops from one energy level to another, a transition sequence is produced. The terms in the transition sequence consist of the differences between the energy at the higher and lower levels.

EXAMPLE

Hydrogen Spectrum

By treating angular momentum in quantum terms, we can determine the atomic spectrum of the hydrogen atom. Using quantized angular momentum and Newton's Law, derive the appropriate atomic spectrum.

SOLUTION

We begin by writing the angular momentum. In quantized form, only certain values are allowed, and so the angular momentum is expressed in terms of n, an integer corresponding to the particular state of the atom, and h (Planck's constant). We then have

$$mvr = \frac{nh}{2\pi} = n\hbar$$

where m is the mass of the electron, v is its speed, and r is the radius of its orbit. The energy of the electron is given by

$$E = \frac{1}{2}mv^2 - \frac{ke^2}{r}$$

where k is a constant and e is the elementary charge corresponding to the charge of a proton. Here the first energy term is the kinetic energy and the second is the result of the positive nucleus attracting the negative electron. From Newton's Law, we obtain

$$\frac{ke^2}{r^2} = \frac{mv^2}{r}.$$

Thus, for the first term we have

$$\frac{1}{2}mv^2 = \frac{ke^2}{2r}.$$

which implies that the energy equation simplifies to

$$E = -\frac{ke^2}{2r}.$$

Combining Newton's Law and the quantization of angular momentum, we obtain

$$\frac{ke^2}{r^2} = \frac{mv^2}{r} \implies \frac{ke^2}{m} = v^2 r$$

and

$$\frac{n^2 \hbar^2}{m^2} = v^2 r^2.$$

Thus,

$$r = \frac{n^2 \hbar^2}{ke^2 m} = n^2 a_o$$

where a_o is called the Bohr radius. The energy is given by

$$E_n = -\frac{ke^2}{2r} = -\frac{k^2 e^4 m}{2\hbar^2 n^2}.$$

The energy of light emitted from an excited hydrogen atom will then be the difference in energy between two different energy levels.

The Lyman Series (discovered by Theodore Lyman in 1906) is a sequence of energy levels resulting from a transition from some higher level to the lowest level, ($n = 1$). The lowest energy level possible in this series is the transition from $n = 2$ to $n = 1$. An infinite sequence of transitions is possible by taking the difference in energy between different levels. Expressed as a wavelength, we obtain

$$\frac{1}{\lambda} = \frac{mk^2 e^4}{4\pi c\hbar^3 [(1/n_f^2) - (1/n_i^2)]}$$

$$\approx 1.097 \times 10^7 \left(\frac{1}{n_f^2} - \frac{1}{n_i^2} \right)$$

where n_i is an excited initial state and n_f is the first state of the atom. In the Lyman Series, the lowest energy emitted corresponds to the longest wavelength, which is 121.5 nanometers. This is in the ultraviolet range. In fact, the entire Lyman Series is ultraviolet.

APPLICATION 9.3 EXERCISES

1. The Balmer Series, first modeled by Johann Jacob Balmer is 1885, consists of the sequence of transition levels, that end at $n = 2$. Determine the number of transitions that are in the visible spectrum (400 nm to 700 nm).

2. Find the range of wavelength of light that can be emitted in the Paschen Series, named for Friederich Paschen who first observed it in 1908, (transition levels end with $n = 3$). Are any of these wavelengths within the visible spectrum?

APPLICATION 9.4: Transforming Difficult Problems

See Section 9.2
Series and Convergence
Calculus, 4th Edition,
Larson/Hostetler/Edwards

The Laplace transform is a fundamental tool in solving electrical circuit problems as well as certain mechanics problems. The basic idea is to convert (transform) a difficult problem into a problem that can be solved more easily.

An RLC electrical circuit with inductance L (in henrys), resistance R (in ohms), capacitance C (in farads), and electromotive force E (in volts) satisfies the following differential equation.

$$\frac{d^2q}{dt^2} + \left(\frac{R}{L}\right)\frac{dq}{dt} + \left(\frac{1}{LC}\right)q = \left(\frac{1}{L}\right)E(t)$$

where q is the charge on the capacitor. (See Figure 9.3.)

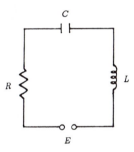

FIGURE 9.3

For example, if $L = 10^{-3}$, $C = 10^{-9}$, $R = 0$, and $E = 120\sin(120\pi t)$, the equation becomes

$$\frac{d^2q}{dt^2} + 10^{12}q(t) = 120{,}000\sin(120\pi t).$$

The Laplace transform can be applied to this equation to convert this problem to a different "transformed" problem. This transformed problem no longer involves derivatives. It can easily be solved using only algebra. For circuit problems involving alternating current, the electromotive force is generally periodic in nature. We consider here the problem of finding the Laplace transform of such a periodic function.

EXAMPLE

Laplace Transforms of Periodic Functions

Let f have period a. Then the Laplace transform of f is defined by

$$\mathcal{L}(f)(s) = \frac{1}{1 - e^{-as}}\int_0^a f(t)e^{-st}\,dt.$$

Let

$$f(t) = \begin{cases} 1, & 0 < t < 1 \\ -1, & 1 < t < 2 \end{cases}$$

have period $a = 2$, as shown in Figure 9.4. Evaluate $\mathcal{L}(f)$ and express the answer as a series involving exponential functions.

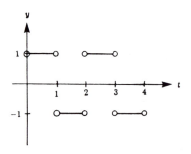

FIGURE 9.4

SOLUTION

We first evaluate

$$\int_0^2 f(t)e^{-st}\,dt = \int_0^1 (1)(e^{-st})\,dt + \int_1^2 (-1)(e^{-st})\,dt$$

$$= \left.\frac{e^{-st}}{-s}\right]_0^1 - \left.\frac{e^{-st}}{-s}\right]_1^2$$

$$= \frac{1-e^{-s}}{s} + \frac{e^{-2s}-e^{-s}}{s}$$

$$= \frac{1-2e^{-s}+e^{-2s}}{s}$$

$$= \frac{(1-e^{-s})^2}{s}.$$

This implies that

$$\mathcal{L}(f)(s) = \frac{1}{1-e^{-2s}}\,\frac{(1-e^{-s})^2}{s}$$

$$= \frac{1}{(1+e^{-s})(1-e^{-s})}\,\frac{(1-e^{-s})^2}{s}$$

$$= \frac{1-e^{-s}}{s(1+e^{-s})}$$

$$= \frac{1-e^{-s}}{s}(1-e^{-s}+e^{-2s}-\cdots)$$

$$= \frac{1}{s}(1-2e^{-s}+2e^{-2s}-2e^{-3s}+\cdots)$$

APPLICATION 9.4 EXERCISES

1. Let

$$f(t) = \begin{cases} 1, & 0 < t < 1 \\ 0, & 1 < t < 2 \end{cases}$$

have period $a = 2$. Evaluate $\mathcal{L}(f)$ and express the answer as a series involving exponential functions.

2. Let $f(t) = \sin t$. Evaluate $\mathcal{L}(f)$. [Hint: Recall that the period of the sine is 2π.]

3. Let $f(t) = \cos t$. Evaluate $\mathcal{L}(f)$.

4. Let $f(t) = \sin at$. Evaluate $\mathcal{L}(f)$.

5. Let $f(t) = \cos at$. Evaluate $\mathcal{L}(f)$.

6. Let $f(t) = e^t$ for $0 < t < 1$, where f has a period of 1. Evaluate $\mathcal{L}(f)$.

7. Let $f(t)$, $f'(t)$, and $f''(t)$ be continuous and have a period of a.

   (a) Show that $\mathcal{L}(f'(t)) = s\mathcal{L}(f) - f(0)$.

   (b) Show that $\mathcal{L}(f''(t)) = s^2\mathcal{L}(f) - sf(0) - f'(0)$.

APPLICATION 9.5: Dipole Fields

> See Section 9.10
> **Taylor and Maclaurin Series**
> Calculus, 4th Edition,
> Larson/Hostetler/Edwards

The fields produced by antennas are often dipole fields where at a given time one half the antenna is charged positive and the other half is charged negative. Magnets also produce dipole fields with one north pole and a corresponding south pole.

A dipole field results from having opposite charges or poles present and then observing the field far from the source. The field, if expressed as a series, will contain terms which are the cubes of the original distances. These terms are the dipole terms. The other terms in the series will be quite small compared to the dipole terms and are usually ignored.

Other arrangements of charges or poles are possible resulting in fields that are not dipole, but perhaps quadra-pole or even higher-order fields. We are only concerned here with dipole fields.

EXAMPLE

Dipoles and the Binomial Theorem

Two uniformly charged rods, one negative and one positive, are placed end to end. (See Figure 9.5.) This should result in an electrical field that looks like a dipole field far away from the charge distribution. Does the field along the line perpendicular to the rods indeed look like a dipole field?

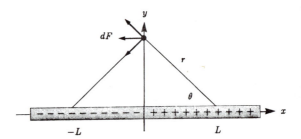

FIGURE 9.5

SOLUTION

The electric field $d\mathbf{E}$ due to a small charge dq is given by Coulomb's Law

$$d\mathbf{E} = \frac{1}{4\pi\epsilon_0 r^2} dq\, \mathbf{r}$$

where ϵ_0 is a constant and $\mathbf{r}$ is a unit vector pointing away from positive charges toward the observation point or toward the negative charges and away from the observation point. We let E represent the magnitude of the electric field $\mathbf{E}$. That part of $d\mathbf{E}$ that is perpendicular to the rod is zero by symmetry. Therefore, that part of $d\mathbf{E}$ that is parallel to the rod is given by

$$dE = \frac{1}{4\pi\epsilon_0} \frac{2 dq \cos\theta}{x^2 + y^2}$$

$$= \frac{1}{4\pi\epsilon_0} \frac{2\lambda\, dx}{x^2 + y^2} \frac{x}{\sqrt{x^2 + y^2}}$$

$$= \frac{\lambda}{4\pi\epsilon_0} \frac{2x\, dx}{(x^2 + y^2)^{3/2}}.$$

$dq = \lambda\, dx$, where λ is the charge density, and $\lambda L = Q$ is the total positive charge. Now, integrating from the center of the rod at $x = 0$ to the end of the rod at $x = L$, we obtain

$$E = \frac{\lambda}{\epsilon_0 4\pi} \int_0^L \frac{2x\, dx}{(x^2 + y^2)^{3/2}}$$

$$= \frac{-2\lambda}{4\pi\epsilon_0 (x^2 + y^2)^{1/2}} \Bigg]_0^L$$

$$= \frac{2\lambda}{4\pi\epsilon_0} \left(\frac{1}{y} - \frac{1}{\sqrt{L^2 + y^2}} \right).$$

Far away from the rod we can write $\sqrt{L^2 + y^2}$ as $y\sqrt{1 + (L/y)^2}$ which yields

$$E = \frac{2\lambda}{4\pi\epsilon_0} \frac{1}{y} \left[1 - 1 + \frac{1}{2}\left(\frac{L}{y}\right)^2 - \frac{(3/2)(1/2)}{2!}\left(\frac{L^2}{y^2}\right)^2 + \cdots \right].$$

Far from the charged object where L/y is small, we need only keep the first non-zero term. Hence,

$$E = \frac{2\lambda}{4\pi\epsilon_0} \frac{1}{y} \frac{1}{2} \frac{L^2}{y^2}$$

$$= \frac{2\lambda L^2}{4\pi\epsilon_0 y^3}.$$

Let $\lambda L = Q$ so we have

$$E = \frac{2QL}{r\pi\epsilon_0} \frac{1}{y^3}$$

where $2QL$ is the dipole moment. We now have the expected dependence of one divided by the cube of the distance which is associated with dipole fields.

APPLICATION 9.5 EXERCISES

1. Find the electric field along the axis of the rods in the example and then find the approximation for the electric field far from the rod. Does the field look like a dipole field?

2. Two rings, each of radius R, are charged with positive and negative charges Q, respectively. The center of the rings are at $(0, 0, \pm a)$. The rings are parallel to the x-y plane. Find the electric field along the z-axis. Does the field look like a dipole field?

3. Two disks, each of radius R, are charged with positive and negative charges Q, respectively. The center of the disks are at $(0, 0, \pm a)$. The disks are parallel to the x-y plane. Find the electric field along the z-axis. Does the field look like a dipole field?

APPLICATION 9.6: The Pendulum

See Section 9.10
Taylor and Maclaurin Series
Calculus, 4th Edition,
Larson/Hostetler/Edwards

The path of a pendulum is dependent on the angle, θ, from the vertical at which it is released, and upon the length, L, of the pendulum. Newton's laws allow us to describe the relationship between θ and L with respect to time as

$$\frac{d^2\theta}{dt^2} + \frac{g}{L}\sin\theta = 0.$$

But what is θ in terms of t? One approach to the solution of this equation is to use a Taylor Series. Another approach, using the same idea of approximation, is to solve the equation in which $\sin\theta$ is approximated by θ. (Is this a reasonable approximation?) By comparing results of both methods, we can achieve an even better understanding of the actual motion.

EXAMPLE

A Pendulum Problem

A pendulum of length two feet is released from rest at an angle $\theta = \pi/6$. (See Figure 9.6)

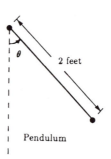

2 feet

θ

Pendulum

FIGURE 9.6

(a) Find the Taylor series solution for θ as a function of the time t. Assume the equation of motion

$$\frac{d^2\theta}{dt^2} + \frac{g}{L}\sin\theta = 0.$$

(b) Approximate the solution by replacing $\sin\theta$ by θ. Compare the two results.

SOLUTION

(a) To use Taylor series, we express $\theta(t)$ as an infinite series. Since $g = 32$ and $L = 2$, we need to solve

$$\frac{d^2\theta}{dt^2} + 16\sin\theta = 0$$

with initial conditions $\theta(0) = \pi/6$ and $\theta'(0) = 0$. We assume that

$$\theta(t) = \sum_{n=1}^{\infty} a_n t^n, \qquad \text{where } a_n = \frac{\theta^{(n)}(0)}{n!}.$$

Note that

$$a_0 = \frac{\theta(0)}{0!} = \frac{\pi}{6} \qquad \text{and} \qquad a_1 = \frac{\theta'(0)}{1!} = 0.$$

We solve for the other coefficients as follows. To solve for a_2, we use the fact that

$$\theta''(t) = -16\sin\theta$$

and conclude that

$$a_2 = \frac{\theta''(0)}{2!} = \frac{-16\sin(\pi/6)}{2!} = -4.$$

Similarly, to solve for a_3, we use the fact that

$$\theta'''(t) = -16\cos\theta\,\frac{d\theta}{dt}$$

and conclude that

$$a_3 = \frac{\theta'''(0)}{3!} = 0.$$

Finally, we use the fact that

$$\theta^{(4)}(t) = 16\sin\theta\left(\frac{d\theta}{dt}\right)^2 - 16\cos\theta\left(\frac{d^2\theta}{dt^2}\right)$$

and conclude that

$$a_4 = \frac{\theta^{(4)}(0)}{4!} = \frac{0 - 16\left(\cos\frac{\pi}{6}\right)\theta''(0)}{4!} = \frac{8\sqrt{3}}{3}.$$

Therefore,

$$\theta(t) = \frac{\pi}{6} - 4t^2 + \frac{8\sqrt{3}}{3}t^4 - \cdots$$

which represents the Taylor series for θ in terms of t.

(b) We can also approximate the solution by replacing $\sin \theta$ by θ in the original pendulum equation. The new equation

$$\frac{d^2\theta}{dt^2} + 16\theta = 0$$

with initial conditions $\theta(0) = \pi/6$ and $\theta'(0) = 0$ has as its solution

$$\theta = \frac{\pi}{6} \cos 4t = \frac{\pi}{6}\left(1 - 8t^2 + \frac{256t^4}{4!} - \cdots\right).$$

We notice that the series in the two solutions have similar forms. The leading term in each series is the same. Moreover, both series have alternating signs and involve only even powers of t.

APPLICATION 9.6 EXERCISES

1. Use a Taylor series to solve the problem $\dfrac{d^2\theta}{dt^2} + \dfrac{g}{L}\sin\theta = 0$ with initial conditions $\theta(0) = \dfrac{\pi}{12}$ and $\theta'(0) = 0$ for $L = 2$ feet.

2. Solve Exercise 1 with initial conditions $\theta(0) = \dfrac{\pi}{12}$ and $\theta'(0) = 2$.

3. Solve the equation $y'' + xy' + y = 0$ with initial conditions $y(0) = 1$ and $y'(0) = 1$.

4. Solve the equation $y'' + xy' + x^2y = 0$ with initial conditions $y(0) = 1$ and $y'(0) = 2$.

5. Solve the equation $x^2y'' + xy' + (x^2 - 1)y = 0$ with initial conditions $y(1) = 1$ and $y'(1) = 2$.

10 CONIC SECTIONS

APPLICATION 10.1 Big Bridges

APPLICATION 10.1: Big Bridges

> See Section 10.1
> Parabolas
> Calculus, 4th Edition,
> Larson/Hostetler/Edwards

The modern bridges of the world can be categorized by the type of construction: concrete-arch, continuous-truss, cable-stayed, steel-arch, cantilever, and suspension. Of these, the type that allows the longest main span is the suspension bridge. The longest suspension bridge in the world, completed in Great Britain in 1981, is the Humber Bridge whose main span is 4626 feet. Three U.S. bridges follow close in ranking: the Verrazano Narrows Bridge (4260 feet, 1964), the Golden Gate Bridge (4200 feet, 1937), and the Mackinac Straits Bridge (3800 feet, 1957).

By way of contrast, the longest cantilever bridge (Quebec Railway, Canada, 1917), is 1800 feet long in its main span. The longest steel-arch bridge (New River Bridge, West Virginia, 1977), is 1700 feet long. The longest continuous-truss bridge (Astoria Bridge of Oregon, 1966), spans a maximum of 1232 feet.

EXAMPLE

The Golden Gate Bridge

The Golden Gate Bridge is a suspension bridge whose main span is 4200 feet long and whose support towers rise 526 feet above the roadway. If the lowest part of the cable is 6 feet above the roadway, find the equation of the cable. (See Figure 10.1.) We assume that the bridge supports a horizontally uniform load.

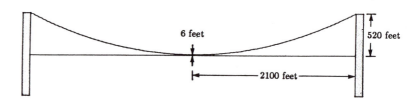

FIGURE 10.1

SOLUTION

We divide the solution into two parts. In the first part we show that the equation of the cable is a parabola. We find the equation of the cable for the Golden Gate Bridge in the second part.

(a) Let the curve be given by $y = f(x)$. (See Figure 10.2.)

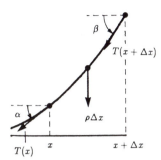

FIGURE 10.2

We first establish an equation involving the forces acting on a segment of the cable. In order for the cable to be in equilibrium, the sum of horizontal and vertical components of the forces on the segment of the cable must be zero. Thus,

$$T(x) \cos \alpha = T(x + \Delta x) \cos \beta = T_0.$$

$T(x)$ and $T(x + \Delta x)$ represent the magnitude of the forces at x and $x + \Delta x$, respectively, and α and β represent the angles at which these forces act with respect to the horizontal. We set T_0 equal to the value in common. We assume that the weight of the segment of the cable is $\rho \Delta x$ where ρ is a constant. So,

$$T(x + \Delta x) \sin \beta - T(x) \sin \alpha - \rho \Delta x = 0.$$

We next obtain a differential equation describing the curve. We substitute the values

$$T(x) = \frac{T_0}{\cos \alpha} \quad \text{and} \quad T(x + \Delta x) = \frac{T_0}{\cos \beta}$$

into the second equation to obtain

$$T_0 \tan \beta - T_0 \tan \alpha = \rho \Delta x.$$

Since

$$f'(x) = \tan \alpha \quad \text{and} \quad f'(x + \Delta x) = \tan \beta,$$

we substitute in the third equation to obtain

$$T_0 f'(x + \Delta x) - T_0 f'(x) = \rho \Delta x.$$

Thus,

$$\frac{f'(x + \Delta x) - f'(x)}{\Delta x} = \frac{\rho}{T_0}.$$

Letting $\Delta x \to 0$, we get

$$f''(x) = \frac{\rho}{T_0} = C.$$

We now integrate twice to obtain

$$y = f(x) = C \frac{x^2}{2} + C_1 x + C_2.$$

Thus, the cable of the suspension bridge is in the shape of a parabola.

(b) We place the vertex of the parabola at the origin and let

$$y = ax^2 + bx + c.$$

(See Figure 10.3.) Substituting the points $(0, 0)$, $(2100, 520)$, and $(-2100, 520)$ into the equation, we obtain $c = 0$, $b = 0$, and $a = 520/(2100)^2$. Thus, the equation is

$$y = \frac{520}{(2100)^2} x^2.$$

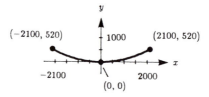

FIGURE 10.3

APPLICATION 10.1 EXERCISES

1. The George Washington Bridge is a suspension bridge whose main span is 3500 feet long and whose support towers rise 377 feet above the roadway. Find the equation of the cable. You may assume that the cable touches midway between the towers.

2. Find the length of a cable for the George Washington Bridge.

3. The Oakland Bay Bridge, damaged in the 1989 earthquake, is a suspension bridge whose main support span is approximately 2300 feet long and whose towers rise approximately 500 feet above the water. If the roadway is an unknown c feet above the water, find the equation of the cable in terms of c. You may assume that the cable touches midway between the towers.

4. A bridge consists of two identical spans, as shown in the figure. Assume that each span is 2000 feet long and that the support towers rise 305 feet above the roadway. If the lowest part of the cable is 5 feet above the roadway, find an equation valid for both parts of the cable.

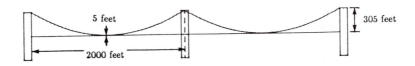

FIGURE FOR EXERCISE 4

11

PLANE CURVES, PARAMETRIC EQUATIONS, AND POLAR COORDINATES

APPLICATION 11.1   Sun, Moon, and Earth
APPLICATION 11.2   The Rotary Engine

APPLICATION 11.1: Sun, Moon, and Earth

> See Section 11.2
> **Parametric Equations and Calculus**
> **Calculus, 4th Edition,**
> **Larson/Hostetler/Edwards**

Newton's Universal Law of Gravitation states that the gravitational force is proportional to the mass of the attracting body and varies with the inverse of the square of the distance between the bodies. We know that the Earth's orbit is nearly circular, with a radius of 93 million miles around the Sun; the Moon's orbit is also nearly circular, with a radius of 240,000 miles around the Earth.

By indirect measurement, we know the mass of the Sun is 2.0×10^{30} kilograms and the mass of the Earth is 5.98×10^{24} kilograms. We assume that the angular velocity of the Moon relative to the Earth is constant. Many interesting relationships can be developed based on just those assumptions; specifically, we can investigate the motion of the Moon to see what most influences its path.

EXAMPLE

Lunar Orbits

The Moon's path around the Earth is such that its position is the same every twenty-eight days. One might believe that the Moon travels around the Earth in loops as the Earth travels around the Sun. This idea is consistent with the idea that the Moon orbits about the Earth and the Earth orbits about the Sun. Which body, the Sun or the Earth, exerts more gravitational pull on the Moon, and therefore dominates its motion?

SOLUTION

We can examine the gravitational attraction between the Moon and the Sun, and between the Moon and the Earth. To see which body is dominant, we compare

$$\frac{GM_S}{r_S{}^2}$$

to

$$\frac{GM_E}{r_E{}^2}$$

where M_S and r_S, M_E and r_E are the mass and radius for the Sun and Earth, respectively. G is the universal gravitational constant whose value is $6.6720 \times 10^{-11} \mathrm{N} \cdot \mathrm{m}^2/\mathrm{kg}^2$. (Because we are interested in relative values, we do not need to actually substitute this value.)

We see that the value for the Sun is 2.31×10^{14}, whereas for the Earth the value is 1.03×10^{14}. The gravitational attraction of the Sun is therefore more than twice that of the Earth.

APPLICATION 11.1 EXERCISES

1. Find the velocity and acceleration vectors for the Moon relative to the Sun. [*Hint:* Place the Sun at the origin and assume that the apparent angular speed of the Moon around the Earth is constant.]

2. Show that the velocity vector and acceleration vector are never parallel. [*Hint:* Use the results of Exercise 1 and the cross products.]

3. The curvature of a path is given by

$$K = \frac{\|\mathbf{v} \times \mathbf{a}\|}{\|\mathbf{v}\|^3}.$$

Show that the curvature of the Moon's orbit is never zero. Interpret the physical meaning of this regarding the orbit of the Moon.

4. Determine whether there is a range of orbital radii within which a satellite will not fall toward the Sun. Find the range of such radii. [*Hint:* This would require the curvature to be zero at some point in the orbit.]

5. Determine the periods associated with the range of radii in which the curvature is never zero and in which the curvature may be zero.

APPLICATION 11.2: The Rotary Engine

> See Section 11.3
> **Polar Coordinates and Polar Graphs**
> **Calculus, 4th Edition,**
> **Larson/Hostetler/Edwards**

The rotary engine was made popular by Felix Wankel in 1954. (See Figure 11.1.) It features a rotor, which is basically a modified equilateral triangle. The rotor moves around a cavity which, in two dimensions, is epitrochoid in shape (peanut-shaped). The shape of the epitrochoid is given parametrically by

$$x = R \cos \theta + b \cos 3\theta$$

and

$$y = R \sin \theta + b \sin 3\theta$$

where b is the distance from the center of a smaller circle to a point along the radius of a smaller circle. Note that $0 < b < R/6$. (See Figure 11.2.) The epitrochoid is then generated by having the smaller circle of radius $R/3$ roll around a fixed circle of radius $2R/3$. The point $p(\theta) = (x, y)$ is a point on the rolling circle located a distance b from the center of the rolling circle.

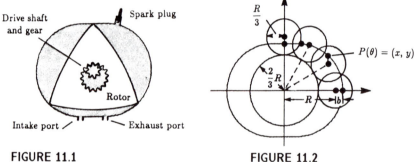

FIGURE 11.1   FIGURE 11.2

The centroid of the rotor traces out a circle about the origin with radius b. In order to transfer the motion of the rotor, a circular hole with gear teeth in the rotor is in contact with a drive gear. The radius of the hole is $2b$ and the drive gear has a radius of b. As the rotor goes around, the centroid of the rotor is opposite the contact point of the rotor with the drive gear.

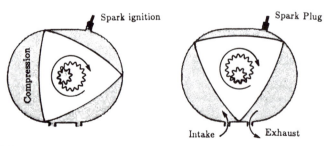

FIGURE 11.3

The engine cycle is illustrated in Figure 11.3. All the phases that occur in a traditional piston engine occur in the rotary engine, but they do so nearly simultaneously. Thus, one rotor of a rotary engine behaves much as a three-cylinder reciprocating engine.

EXAMPLE
━━━━━

Rotor Wipers

Show that the rotor wipers remain in contact with the chamber wall.

SOLUTION

The wipers actually correspond to vertices of an equilateral triangle. We will establish that the distance between an arbitrary point on the epitrochoid and another point corresponding to an increase in the parameter θ by $2\pi/3$ are independent of θ. There will therefore be three such distances, returning to the original point. We will work with the distance squared since this simplifies the work. The distance squared, which we will call M, is given by

$$M = (x_2 - x_1)^2 + (y_2 - y_1)^2$$

$$= \left[R\cos\left(\frac{2\pi}{3} + \theta\right) + b\cos 3\left(\frac{2\pi}{3} + \theta\right) - R\cos\theta - b\cos 3\theta \right]^2$$

$$+ \left[R\sin\left(\frac{2\pi}{3} + \theta\right) + b\sin 3\left(\frac{2\pi}{3} + \theta\right) - R\sin\theta - b\sin 3\theta \right]^2$$

We immediately notice that the terms containing b drop out since $3[(2\pi/3) + \theta] = 2\pi + 3\theta$ and sine and cosine are periodic with a period of 2π. This leaves

$$M = R^2\left[\cos\left(\frac{2\pi}{3} + \theta\right) - \cos\theta\right]^2 + R^2\left[\sin\left(\frac{2\pi}{3} + \theta\right) - \sin\theta\right]^2$$

$$= R^2\left[\cos^2\left(\frac{2\pi}{3} + \theta\right) - 2\cos\left(\frac{2\pi}{3} + \theta\right)\cos\theta + \cos^2\theta\right.$$

$$\left. + \sin^2\left(\frac{2\pi}{3} + \theta\right) - 2\sin\left(\frac{2\pi}{3} + \theta\right)\sin\theta + \sin^2\theta\right]$$

$$= R^2\left[2 - 2\cos\left(\frac{2\pi}{3} + \theta - \theta\right)\right]$$

$$= R^2\left(2 - 2\cos\frac{2\pi}{3}\right)$$

$$= 3R^2.$$

Thus, the distance between vertices is a constant $\sqrt{3}\,R$.

━━━━━━━━━━━━━━━━━━━━━━━━━━━━━━━

APPLICATION 11.2 EXERCISES

1. (a) Show that the centroid of the rotor traces out a circle of radius b centered at the origin.
   (b) Show that the centroid is traced three times for each complete rotation of the rotor.
   (c) Determine the number of revolutions for the drive gear for each turn of the rotor.
2. Assume that the legs of the triangle are replaced with arcs of circles which are centered at the opposite vertices. For the rotor to fit, b must be less than or equal to $(2 - \sqrt{3})R/2$. Verify this result. The advantage of this shape is that the diameter of the rotor is a constant.

3. The compression ratio is the ratio of the largest volume contained within the compression part of the cycle to the smallest volume contained within this part of the cycle. Assuming that $b = (2-\sqrt{3})R/2$, what is the compression ratio for the rotary engine? *Hint:* Consider finding the area between the rotor and cavity for any given angle and then find the critical points. Also recall that the adjacent vertices of the rotor are $2\pi/3$ radians apart.

4. Current rotary engines used in one imported car model have a compression ratio of 9.4 : 1. Determine the value of b which would result in this compression ratio.

12

VECTORS AND THE GEOMETRY OF SPACE

APPLICATION 12.1: Boomerangs

> See Section 12.4
> **The Cross Product of Two Vectors in Space**
> **Calculus, 4th Edition,**
> **Larson/Hostetler/Edwards**

We all know that a boomerang, when properly thrown, will return to the point from which it was originally thrown. Can a quantitative explanation for boomerang behavior be developed?

FIGURE 12.1

Understanding the behavior of a boomerang requires a study of its shape. (See Figure 12.1.) The cross section of a boomerang is like the cross section of an airplane wing, with one flat side and one curved side. When the air rushes past the boomerang, it movers faster over the curved side. By Bernoulli's Principle, this causes reduced pressure (lift) on the curved side.

As a boomerang spins and moves through the air, the lift is continually changing, depending on the current position of the boomerang. We shall consider two extreme positions, as shown in Figure 12.2.

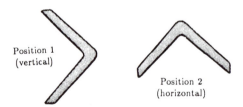

FIGURE 12.2

When the boomerang is in Position 1, there is a lift toward the curved side of the boomerang. In addition, the upper wing is moving faster relative to the stationary air, so the lift is even greater. This causes a torque on the boomerang. A torque results in a change in angular momentum, which causes the plane in which the boomerang is rotating to change. Considering the torque $\vec{\tau}$ about the center of mass, $\vec{\tau} = \mathbf{r} \times \mathbf{F}$, the net torque, in general, points toward the direction opposite to the direction of motion. Angular momentum, which essentially points along the axis of rotation, changes according to Newton's Law, $\vec{\tau} = d\mathbf{L}/dt$. Thus, the axis of rotation rotates back, causing the boomerang to curve back.

When the boomerang is in Position 2, the front wing gets lift but the back wing is in the "shadow" of the front wing and does not get much lift. For reasons similar to those given above, the torque points up. The boomerang will react by rotating the angular momentum vector upward, causing the plane of rotation to move from the initial vertical position to one that is more horizontal.

EXAMPLE

Calculating Lift on a Boomerang

Let us consider some assumptions that we will use in quantitatively developing our model boomerang. We assume that Bernoulli's Principle is appropriate. This principle states that

$$P + \frac{\rho v^2}{2} + \rho g h = \text{constant}$$

where P is the air pressure on the moving surface (pressure is always perpendicular to the surface), ρ is the density of air (which is 1.29 kg/m^3), g is the acceleration due to gravity, v is the speed of the air passing over the wing surface, and h is the height of the area being considered. We will assume that the density of air is constant, and that the air does not compress due to the boomerang interacting with it. At the speeds we consider in this calculation, this assumption is quite acceptable.

Some estimates of speeds are in order for our boomerang model. We will assume a linear speed of 10 meters/second and a rotational speed of 10 radians/second. Our model boomerang has dimensions, as illustrated in Figure 12.3.

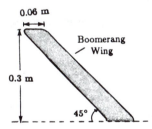

FIGURE 12.3

The path over the curved surface is slightly longer than the flat surface. We will assume that air must travel 5% faster over the curved surface to reach the back edge of the wing than when traveling over the flat surface. Since we are only approximating, we assume that the average linear speed across the wing as being the sum of the linear speed of the center of mass and the average speed due to rotation. The speed is, therefore

$$10 \text{ m/sec} + (0.15 \text{ m})(10 \text{ rad/sec}) = 11.5 \text{ m/sec}.$$

What then, is a reasonable estimate of the lift on the upper wing of boomerang when it is in a vertical position?

SOLUTION

Starting with Bernoulli's Principle

$$P + \frac{\rho v^2}{2} + \rho g h = \text{constant}$$

we observe that the last term, $\rho g h$ can be ignored in our calculations because this term will not contribute to the result. The difference in height for the curved and corresponding flat surface is very small when the boomerang is horizontal. Moreover, in our present case, when the boomerang is vertical, the difference in height for the curved and corresponding flat surface is zero. We can use all the previously assumed values to calculate the difference in pressure between the curved and flat surface of the upper wing of the vertical boomerang.

$$\Delta P = P_1 - P_2$$
$$= \frac{1}{2}(1.29)(11.5)^2(1.05^2 - 1)$$
$$= 8.7 \text{ N/m}^2.$$

The net force due to this pressure difference is

$$F = (\text{area})\Delta P$$
$$= (0.3)(0.06)\sin 45° \Delta P$$
$$= 0.111 \text{ N}.$$

APPLICATION 12.1 EXERCISES

1. Find the lift on the lower wing for our model boomerang when it is in the vertical position.

2. Find the lift of the front wing when the boomerang is in the horizontal position.

3. Find the net torque on the center of mass for the boomerang in the vertical position, consistent with the assumptions already made.

4. Find the net torque on the center of mass for the boomerang in the horizontal position assuming the front wing provides lift but the back wing does not because it is in the shadow of the front wing.

APPLICATION 12.2: Electromagnetism Experiments

> **See Section 12.4**
> **The Cross Product of Two Vectors in Space**
> **Calculus, 4th Edition,**
> **Larson/Hostetler/Edwards**

In electromagnetism, the force felt by a charged particle moving in an electromagnetic field is called the Lorentz Force. It is given by

$$\mathbf{F} = q\mathbf{E} + (q\mathbf{v} \times \mathbf{B})$$

where $\mathbf{F}$ is the force on the particle, $\mathbf{E}$ is the electric field, $\mathbf{v}$ is the velocity of the particle, and $\mathbf{B}$ is the magnetic field. The potential difference (voltage) is measured from plus to minus. So, if positive charges line up on one side of region and negative charges line up on the other side, the result will be a "+ to −" voltage.

Using this information, we can determine whether positive or negative particles are carrying the electric current. The determination of the charge of carriers was first discovered by Edwin Hall in 1879. The method used to determine the charge of carriers is called the Hall Effect.

EXAMPLE

Hall Effect

Use the Hall Effect and the Lorentz Force to determine the charge of electrical carriers.

SOLUTION

We begin by assuming that the carriers are positive. We also assume that there is a current flowing across the page from left to right, as illustrated in Figure 12.4.

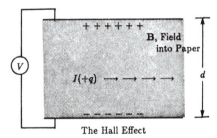

The Hall Effect

FIGURE 12.4

In equilibrium,

$$\mathbf{F} = \mathbf{0} = q\mathbf{E} + (q\mathbf{V} \times \mathbf{B})$$

so $E = vB$. E, v, and B represent the corresponding magnitudes of $\mathbf{E}$, $\mathbf{V}$, and $\mathbf{B}$. In a uniform electrical field $E = vd$, where v is the potential difference between the two edges of the conductors, we can write $v/d = vB$ where v is the velocity of the moving charges, which can be shown to be $v = I/(nqA)$. Here, n is the number of carriers per unit volume and A is the cross-sectional area of the conductor, which is also given by td where t is the thickness. Therefore,

$$v = vBd = \frac{IBd}{nqA} = \frac{IBd}{nqtd} = \frac{BI}{nqt}.$$

One of the uses of Hall probes is to measure magnetic fields.

APPLICATION 12.2 EXERCISES

1. Show that the Hall voltage will be reversed if the carriers are negative instead of positive. Remember that this requires the carriers to flow in the opposite direction, that is, from negative to positive.

2. The Thomson experiment, named after J. J. Thomson (1856–1940), was used to determine the ratio of an electric charge e to a mass m. A cathode ray tube is used in the experiment (See figure.) This tube first sends electrons from a cathode, or hot metal element, through a potential difference to achieve a velocity. Next, the electrons are passed through a slit to provide a thin stream of electrons. The rest of the tube consists of parallel plates, which can have a potential difference placed on them, and a coil, which can provide a magnetic field perpendicular to the electric field on the parallel plates. The procedure is to first let the electrons strike the screen. Next apply the electric field. This causes the spot produced by the electron beam to be deflected. Finally, a magnetic field is applied so that the beam is no longer deflected.

   (a) Show that the cathode ray tube deflection of the electron beam with no magnetic field is given by
   $$y = \frac{eEL^2}{2mv^2}.$$

   (b) Show that when the beam in the cathode ray tube is restored to the original non-deflected position, that $E = vB$.

   (c) Use the results of parts (a) and (b) to show that the ratio e/m is given by
   $$\frac{e}{m} = \frac{2yE}{(BL)^2}.$$

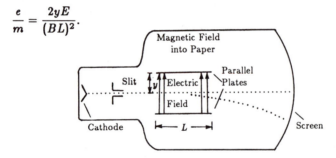

The Thomson Experiment

FIGURE FOR EXERCISE 2

APPLICATION 12.3: The Biot-Savart Law

> **See Section 12.4**
> **The Cross Product of Two Vectors in Space**
> **Calculus, 4th Edition,**
> **Larson/Hostetler/Edwards**

The Biot-Savart Law is used to calculate magnetic fields **B** given a current i flowing in some region. The differential form of this law is

$$d\mathbf{B} = \frac{\mu_0 i}{4\pi r^2}(d\mathbf{l} \times \hat{\mathbf{r}})$$

where μ_0 is a constant, $d\mathbf{l}$ is an element of wire with current i flowing in the direction inducted by $d\mathbf{l}$, and $\hat{\mathbf{r}}$ is a unit vector pointing from $d\mathbf{l}$ toward the point where the magnetic field is being determined.

EXAMPLE

The Biot-Savart Law

In problems with a high degree of symmetry, this complicated-looking expression is often quite simple. We shall use the example of a wire loop, shown in Figure 12.5, to find the magnetic field along the axis of the loop which is assumed to be carrying a current i.

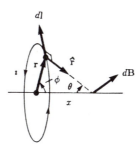

FIGURE 12.5

SOLUTION

We see by the right-hand rule for cross products, by the direction of $d\mathbf{B}$, and by symmetry, that the only nonzero term is the component along the axis. We call this component $dB_{||}$. Since $d\mathbf{l}$ and $\hat{\mathbf{r}}$ are perpendicular, we have

$$dB_{||} = \frac{\mu_0 i}{4\pi}\left(\frac{dl\sin\theta}{r^2}\right)$$

$$B_{||} = \frac{\mu_0 i}{4\pi}\int_0^{2\pi}\frac{r\,d\phi}{x^2+y^2}\sin\theta$$

$$= \frac{2\pi\mu_0}{4\pi}\left(\frac{ir^2}{(x^2+y^2)^{3/2}}\right).$$

Notice that this can be written as

$$B_{||} = \frac{\mu_0}{2\pi}\left(\frac{i\pi r^2}{(x^2+y^2)^{3/2}}\right) = \frac{\mu_0}{2\pi}\left(\frac{M}{(x^2+y^2)^{3/2}}\right)$$

where $M = i\pi r^2 = iA$ is the magnetic dipole moment.

APPLICATION 12.3 EXERCISES

1. Use the Biot-Savart Law to determine the magnetic field a distance D away from a finite wire of length L, with current i.

2. Repeat Exercise 1 for an infinite length wire.

3. Find the magnetic field along the axis of a square loop of wire of length L, carrying current i, x units from the plane of the loop.

4. Refer to the accompanying figure. Determine the magnetic field at the center of the two semi-circles shown in parts (a) and (b). Assume that the current is i.

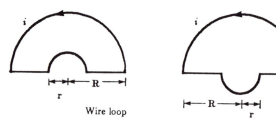

FIGURE FOR EXERCISE 4

13

VECTOR-VALUED FUNCTIONS

APPLICATION 13.1: Interception Problem

> See Section **13.3**
> **Velocity and Acceleration**
> **Calculus, 4th Edition,**
> **Larson/Hostetler/Edwards**

EXAMPLE

Intercepting a Bomb

A bomber is flying horizontally at an altitude of 3200 feet with a velocity of 400 feet/second when it releases a bomb. (See Figure 13.1.) A projectile is launched five seconds later from a cannon at a site facing the bomber and 5000 feet from the point beneath the original position of the bomber. If the projectile is to intercept the bomb at an altitude of 1600 feet, determine the initial speed and angle of inclination for the projectile. Ignore air resistance.

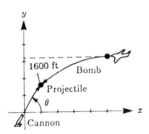

FIGURE 13.1

SOLUTION

To begin, we let v_0 be the initial speed and let θ be the angle of inclination for the projectile at launch. Note in Figure 13.1 that the launch site for the projectile is at the origin. Since the horizontal component of the velocity for the bomb is 400 feet/second, the components of the bomb are

$$x_b = x_0 - (v_0 \cos 0)t = 5000 - 400t$$

and

$$y_b = y_0 + (v_0 \sin 0)t - \frac{1}{2}gt^2 = 3200 - 16t^2.$$

The position vector for the bomb is thus

$$\mathbf{r}_b(t) = x_b\,\mathbf{i} + y_b\,\mathbf{j} = (5000 - 400t)\,\mathbf{i} + (3200 - 16t^2)\,\mathbf{j}.$$

For the projectile, the initial horizontal component of the velocity is $v_0 \cos \theta$ and the initial vertical component of the velocity is $v_0 \sin \theta$. Since the projectile is launched 5 seconds after the bomb is released, the components for the projectile are

$$x_p = (v_0 \cos \theta)(t - 5)$$

$$y_p = (v_0 \sin \theta)(t - 5) - 16(t - 5)^2.$$

The position vector for the projectile is thus

$$\mathbf{r}_p(t) = x_p\,\mathbf{i} + y_p\,\mathbf{j} = (v_0 \cos \theta)(t - 5)\,\mathbf{i} + [(v_0 \sin \theta)(t - 5) - 16(t - 5^2)]\,\mathbf{j}.$$

We set

$$y_b = 1600 = 3200 - 16t^2$$

to obtain $t = 10$ seconds. Thus,

$$x_b = 5000 - 400t = 1000 \text{ feet.}$$

We set $t = 10$, $y_p = 1600$, and $x_p = 1000$ and solve simultaneously for v_0 and θ as follows.

$$v_0(\cos\theta)(5) = 1000$$
$$v_0\cos\theta = 200$$
$$v_0(\sin\theta)(5) - 16(25) = 1600$$
$$v_0\sin\theta = 400$$

Thus,

$$\tan\theta = 2$$
$$\theta \approx 1.107 \text{ radians.}$$

Since $\sin\theta = 2/\sqrt{5}$, we find that

$$v_0 = \frac{400}{\sin\theta} = 200\sqrt{5} \text{ ft/sec.}$$

APPLICATION 13.1 EXERCISES

1. Solve the given example except assume that the horizontal component of the bomb is zero.

2. Solve the given example except assume that the bomber is facing away from the launch site. (See the figure for Exercise 2.)

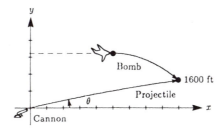

FIGURE FOR EXERCISE 2

3. Consider the given example except assume that the projectile is launched with an initial speed of 800 feet/second but the launch time is unknown. Determine the angle of inclination for the projectile at launch and determine the launch time.

4. Consider the bomber and cannon as in the given example. A missile is launched by the bomber at the same time that the bomb is dropped. If the missile is to destroy the cannon prior to the launch of the projectile, determine the initial velocity and the angle of launch for the missile.

APPLICATION 13.2: Satellite Acceleration

> **See Section 13.3**
> **Velocity and Acceleration**
> **Calculus, 4th Edition,**
> **Larson/Hostetler/Edwards**

As we have seen, polar coordinates are frequently the best means of describing many physical problems. We consider here the problem of finding the radial and angular components of acceleration.

EXAMPLE

Satellite Acceleration

A communications satellite moves in a circular orbit around the Earth at a distance of 42,000 kilometers from the center of the Earth. The angular velocity $d\theta/dt = \omega = \pi/12$ radians per hour is constant. Find the radial and angular components of the acceleration. (See Figure 13.2.)

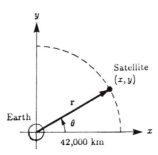

FIGURE 13.2

SOLUTION

Let $\mathbf{r} = x\mathbf{i} + y\mathbf{j}$ be the position vector and let r and θ be the usual polar coordinates. We first show

$$\mathbf{a} = \frac{d^2\mathbf{r}}{dt^2} = \left[\frac{d^2r}{dt^2} - r\left(\frac{d\theta}{dt}\right)^2\right]\mathbf{u}_r + \left[r\frac{d^2\theta}{dt^2} + 2\frac{dr}{dt}\frac{d\theta}{dt}\right]\mathbf{u}_\theta.$$

Since

$$\mathbf{r} = x\mathbf{i} + y\mathbf{j} = r\cos\theta\,\mathbf{i} + r\sin\theta\,\mathbf{j}$$

we have

$$\frac{d\mathbf{r}}{dt} = \left[\frac{dr}{dt}\cos\theta - r\sin\theta\frac{d\theta}{dt}\right]\mathbf{i} + \left[\frac{dr}{dt}\sin\theta + r\cos\theta\frac{d\theta}{dt}\right]\mathbf{j}$$

$$\frac{d^2\mathbf{r}}{dt^2} = \left[\frac{d^2r}{dt^2}\cos\theta - \frac{dr}{dt}\sin\theta\frac{d\theta}{dt} - \frac{dr}{dt}\sin\theta\frac{d\theta}{dt} - r\cos\theta\left(\frac{d\theta}{dt}\right)^2 - r\sin\theta\frac{d^2\theta}{dt^2}\right]\mathbf{i}$$

$$+ \left[\frac{d^2r}{dt^2}\sin\theta + \frac{dr}{dt}\cos\theta\frac{d\theta}{dt} + \frac{dr}{dt}\cos\theta\frac{d\theta}{dt} - r\sin\theta\left(\frac{d\theta}{dt}\right)^2 + r\cos\theta\frac{d^2\theta}{dt^2}\right]\mathbf{j}.$$

The unit vector in the radial direction is given by

$$\mathbf{u}_r = \cos\theta\,\mathbf{i} + \sin\theta\,\mathbf{j}.$$

Thus, the radial component of the acceleration is

$$a_r = \mathbf{a} \cdot \mathbf{u}_r$$

$$= \frac{d^2 r}{dt^2} \cos^2 \theta - 2 \frac{dr}{dt} \cos \theta \sin \theta \frac{d\theta}{dt} - r \cos^2 \theta \left(\frac{d\theta}{dt} \right)^2 - r \cos \theta \sin \theta \frac{d^2 \theta}{dt^2}$$

$$+ \frac{d^2 r}{dt^2} \sin^2 \theta + 2 \frac{dr}{dt} \cos \theta \sin \theta \frac{d\theta}{dt} - r \sin^2 \theta \left(\frac{d\theta}{dt} \right)^2 + r \cos \theta \sin \theta \frac{d^2 \theta}{dt^2}$$

$$= \frac{d^2 r}{dt^2} - r \left(\frac{d\theta}{dt} \right)^2.$$

Using a similar argument, we obtain the angular component of acceleration.

$$a_\theta = r \frac{d^2 \theta}{dt^2} + 2 \frac{dr}{dt} \frac{d\theta}{dt}$$

The acceleration can then be written

$$\mathbf{a} = \frac{d^2 \mathbf{r}}{dt^2} = \left[\frac{d^2 r}{dt^2} - r \left(\frac{d\theta}{dt} \right)^2 \right] \mathbf{u}_r + \left[r \frac{d^2 \theta}{dt^2} + 2 \frac{dr}{dt} \frac{d\theta}{dt} \right] \mathbf{u}_\theta.$$

For the satellite, we have

$$\mathbf{r} = 42000 \cos \left(\frac{\pi}{12} t \right) \mathbf{i} + 42000 \sin \left(\frac{\pi}{12} t \right) \mathbf{j}.$$

Using $r = 42000$, $dr/dt = 0$, $d^2 r/dt^2 = 0$, $d\theta/dt = \pi/12$, and $d^2 \theta/dt^2 = 0$, we obtain

$$\mathbf{a} = -42000 \left(\frac{\pi}{12} \right)^2 \mathbf{u}_r.$$

Thus, the acceleration of the satellite is directed toward the center of the Earth.

APPLICATION 13.2 EXERCISES

1. Show that the angular component of **a** is

$$a_\theta = r \frac{d^2 \theta}{dt^2} + 2 \frac{dr}{dt} \frac{d\theta}{dt}.$$

2. Let $\mathbf{r} = t \cos(\omega t) \mathbf{i} + t \sin(\omega t) \mathbf{j}$. Find the radial and angular components of the acceleration.

3. Kepler's Second Law states that the planets move in orbits so that the rays from the Sun to the planet sweep out equal areas in equal times. Place the Sun at the origin and assume that the acceleration for a planet is $\mathbf{a} = -(c/r^2)\mathbf{u}_r$. Show that

$$\frac{dA}{dt} = \frac{1}{2} r^2 \frac{d\theta}{dt}$$

is a constant.

4. Let $\mathbf{r} = a \cos(\omega t) \mathbf{i} + b \sin(\omega t) \mathbf{j}$. Show that the acceleration is in the direction opposite to $\mathbf{r}(t)$.

5. If the acceleration $\mathbf{a}(t)$ is in the direction opposite to $\mathbf{r}(t)$, show that the path lies in a plane.

6. If $\mathbf{a}(t) = -c\,\mathbf{r}(t)$ where c is a positive constant, show that the path is an ellipse. You may assume the initial conditions $t = 0$, $x = a$, $y = 0$, $dx/dt = 0$, and $dy/dt = 0$.

APPLICATION 13.3: Curvature of a Train Track

> See Section 13.5
> Arc Length and Curvature
> Calculus, 4th Edition,
> Larson/Hostetler/Edwards

Trains are most likely to derail when changing from a straight track to a circular segment of track or vice versa. This is not surprising, for if a train is traveling at a constant speed, v, when on the straight part of the track, it experiences no acceleration. While on the circular part, the train has an acceleration of v^2/r, where r is the radius of curvature and v is the speed of the train. The reciprocal of the radius of curvature is the curvature K, which is given in various forms. Typically,

$$K = \frac{\|\mathbf{v} \times \mathbf{a}\|}{v^3}.$$

There is a famous horseshoe curve at Goldtree, California, just north of San Luis Obispo, along the Southern Pacific right-of-way. It consists of a turn which completely reverses the direction of travel of the train. This horseshoe turn is one of two such turns currently in use in the United States. There is a reverse curve leading into the 1400-foot diameter circular part of the curve. The maximum allowable speed on the curve is 25 miles/hour, or about 37 feet/second.

EXAMPLE

Reducing the Likelihood of Derailment

A straight track is matched to a transition track in the Goldtree horseshoe to reduce the likelihood of a derailment. What would the change in acceleration be on the Goldtree horseshoe turn if the track were an arc of a circle mated to the straight track? (See Figure 13.3.)

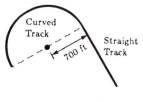

(No Transition Track)

FIGURE 13.3

SOLUTION

The acceleration is given by v^2/r on the circular track, and there is no acceleration on the straight part. Therefore, the change in acceleration is

$$\frac{37^2}{700} = 1.96 \text{ ft/sec}^2.$$

To try to offset this centripetal acceleration, the outer track is four inches higher than the inner track.

APPLICATION 13.3 EXERCISES

1. Show that the expression for curvature,

$$K = \frac{\|\mathbf{v} \times \mathbf{a}\|}{v^3}.$$

reduces to

$$K = \frac{y''}{(1 + [y']^2)^{3/2}}.$$

2. Obtain the lowest-degree polynomial that describes the transition track for a straight track running along the negative x-axis up to $x = 0$ matched up with a circular track of radius r. The circular track segment lies along the circle described by $x^2 + (y - r)^2 = r^2$. Replace the first $\pi/4$ of the arc of the circular track. (See the figure.) [*Hint:* Match the function, derivative, and curvature at each junction.]

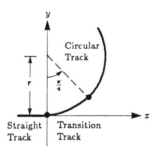

FIGURE FOR EXERCISE 2

3. The transition track connects the circle expressed as $x^2 + y^2 = 700^2$ to the straight track $y = -x + 900$. The junctions occur at $(0, 700)$ and $(900, 0)$. Using the data in the figure and the hint in Exercise 2, find the lowest-degree polynomial describing the transition track.

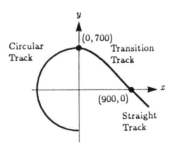

FIGURE FOR EXERCISE 3

14

FUNCTIONS OF SEVERAL VARIABLES

APPLICATION 14.1: The Vibrating String

> See Section 14.5
> Chain Rules for Functions of Several Variables
> Calculus, 4th Edition,
> Larson/Hostetler/Edwards

The motion of a vibrating string can be modeled from the wave equation, a fundamental equation of mathematical physics

$$y_{xx} = \frac{1}{c^2} y_{tt}$$

where $c = \sqrt{T/\rho}$, T is the tension, ρ is the density, and c is the speed of the propagating wave within the string. This partial differential equation can be easily solved by use of the Chain Rule if we know the appropriate boundary and initial conditions. Let's assume that our string is infinite, which eliminates the boundary conditions. We shall also assume that the string has no kinks or corners and that there are no abrupt changes in velocity in adjacent segments of the string. (Corners or abrupt changes in velocity along the string would destroy the string.) These assumptions imply mathematically that the initial displacement function f is twice differentiable and that the initial velocity function g is also differentiable. Let's also assume sufficient differentiability and continuity so that the order of partial differentiation can be interchanged. Under these assumptions we can develop the general solution of the wave equation.

To obtain the general form of our solution, we begin by stating a change in variables.

$$n = x - ct \quad \text{and} \quad s = x + ct$$

Then $y_x = y_n n_x + y_s s_x$. However, $n_x = s_x = 1$, so $y_x = y_n + y_s$ and we have

$$y_{xx} = y_{nn} n_x + y_{ns} s_x + y_{sn} n_x + y_{ss} s_x$$
$$= y_{nn} + 2y_{ns} + y_{ss}.$$

Next, we consider $y_t = y_n n_t + y_s s_t$. If $n_t = -c$ and $s_t = c$, then $y_t = -cy_n + cy_s$ and

$$y_{tt} = -cy_{nn} n_t - cy_{ns} s_t + cy_{sn} n_t + cy_{ss} s_t$$
$$= c^2 (y_{nn} - 2y_{ns} + y_{ss}).$$

Putting this together, we get

$$y_{xx} = \frac{1}{c^2} y_{tt}$$
$$y_{nn} + 2y_{ns} + y_{ss} = \frac{c^2 (y_{nn} - 2y_{ns} + y_{ss})}{c^2}$$
$$4y_{ns} = 0.$$

At first, this looks like it says very little, but it will yield a very powerful solution, D'Alembert's solution, to the wave equation.

By writing $y_{ns} = (y_s)_n = 0$, we know that y_s is constant with respect to n. Thus, the most general form of y_s must be some function of s alone. That is, $y_s = \theta'(s)$, where the prime denotes differentiation with respect to the argument of the function. Integrating with respect to s, we get $y = \theta(s) + \text{constant}$, where the constant could be a function of n. We could, however, have called our mixed partial derivative y_{sn}. Then we would have obtained $y = \phi(n) +$ constant, where the constant could be a function of s. Combining these results, we obtain

$$y = \phi(n) + \theta(s) = \phi(x - ct) + \theta(x + ct).$$

We now have the general form of the solution. However, we must put it together with the initial conditions. At $t = 0$, we have the two initial condition equations:

$$y(x, \; 0) = f(x) = \theta(x) + \phi(x)$$

and

$$y_t(x, \; 0) = g(x) = -c\theta'(x) + c\phi'(x).$$

We can integrate the second initial condition equation to get

$$k + \frac{1}{c} \int_0^x g(v) \, dv = -\theta(x) + \phi(x).$$

The integral here is written as a definite integral with lower limit 0, but there is no loss of generality because we have introduced a constant k, which is arbitrary. Solving for θ and ϕ, we get

$$\theta = \frac{1}{2} f(x) + \frac{1}{2c} \int_0^x g(v) \, dv + k$$

and

$$\phi = \frac{1}{2} f(x) - \frac{1}{2c} \int_0^x g(v) \, dv - k.$$

Our solution is now

$$y(x, \; t) = \theta(x + ct) + \phi(x - ct)$$

$$= \frac{1}{2}[f(x + ct) + f(x - ct)] + \frac{1}{2c} \left[\int_0^{x+ct} g(v) \, dv - \int_0^{x-ct} g(v) \, dv \right] + k - k.$$

Notice that the arbitrary constant k disappears. We can now rewrite the two integrals as one if we reverse the order of integration in the second integral, resulting in

$$y(x, \; t) = \frac{1}{2}[f(x + ct) + f(x - ct)] + \frac{1}{2c} \int_{x-ct}^{x+ct} g(v) \, dv.$$

This is a very compact solution for such a complicated problem. However, we have not yet begun to see the beauty of this solution—this comes from a geometric interpretation.

If we plot a graph of t versus x, commonly called an x-t diagram, we can get a great deal of insight. It would be preferable to have the value of y for each $(x, \; t)$ pair, but since that requires three dimensions and is much more difficult to sketch, we will continue with the two-dimensional interpretation of the x-t diagram. We relax the differentiability conditions on the initial conditions, knowing that the results are not physically possible. The understanding gained is worth the sacrifice.

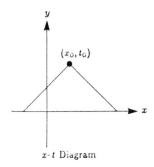

FIGURE 14.1

Consider the x-t diagram in Figure 14.1. The displacement at $(x_0,\ t_0)$ is obtained by drawing the straight lines $x-ct = x_0-ct_0$ and $x+ct = x_0+ct_0$. These lines (called characteristics) are the lines for which the intermediate variables n and s are constant.

The interpretation can now be made. The displacement of the string at the point $(x_0,\ t_0)$ consists of two terms. The first term is simply the average of the initial displacements where the characteristics intersect the x-axis. The second term is the result of the average velocity acting for the t_0. This can be seen to be true by rewriting the second term as

$$\frac{1}{2c}\int_{x_0-ct_0}^{x_0+ct_0} g(v)\,dv = \left[\frac{1}{t_0}\left(\frac{1}{2c}\int_{x_0-ct_0}^{x_0+ct_0} g(v)\,dv\right)\right] t_0$$

where we have integrated the initial velocity over an interval whose length is $2ct_0$ and then divided by that length.

EXAMPLE

Geometric Interpretation

Consider the initial displacement to be a square pulse. (The corresponding string situation is not sufficiently smooth, but we will interpret the x-t diagram for the function nonetheless.) Let the initial velocity be 0. Then we have

$$f(x) = \begin{cases} 0, & |x| > 1 \\ H, & |x| < 1. \end{cases}$$

Sketch the x-t diagram and use it to interpret the string displacement.

SOLUTION

The x-t diagram can be divided into regions as illustrated in Figure 14.2.

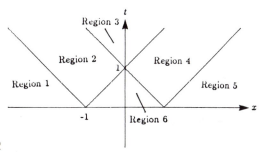

FIGURE 14.2

The characteristics are radiating from $x = -1$ and $x = 1$ along the x-axis, dividing the upper half-plane into six regions. By tracing the characteristics back to the x-axis to obtain the average initial displacements, we obtain the displacement at our point (x_0, t_0). (See Figure 14.3.)

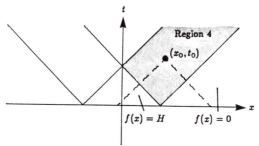

FIGURE 14.3   Propagation of Initial Displacement

We see for Region 4 at our point that the values traced back along the characteristics are H and 0, so the displacement will be

$$\frac{1}{2}(H + 0) = \frac{H}{2}.$$

In a similar way, we can get the values of the displacement of our string for the other regions, as shown in Figure 14.4. The values obtained are written in their corresponding regions.

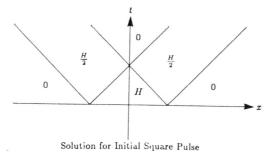

FIGURE 14.4   Solution for Initial Square Pulse

We can now sketch the displacement as a third dimension for various times. This will indicate how the position of the string looks at the times selected. (See Figure 14.5.) Here any displacement from a particular t (horizontal line) is considered a displacement from $y = 0$ for the string.

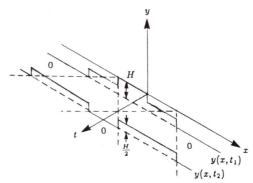

FIGURE 14.5

String Displacement at Different Times t_1 and t_2

APPLICATION 14.1 EXERCISES

1. Let the initial displacement of a long string be zero and let the initial velocity be given by

$$g(x) = \begin{cases} V, & |x| < 1 \\ 0, & |x| > 1. \end{cases}$$

   Although this is physically not possible, we can still interpret the results. Construct an x-t diagram and divide it into regions as in the example. Then obtain expressions for the displacement in each region. Sketch representative displacements for several values of t.

2. An initial displacement that is more realistic than the square pulse used in the example is $f(x) = Ae^{-x^2}$. It is more realistic because it can be differentiated. With no initial velocity, determine the displacement everywhere.

3. A guitar string can be modeled as an infinitely long string with fixed points at $x = 0$ and $x = L$, and with nodes at $x = nL$, where n is an integer. Consider several values of t with $0 < x < L$, and show that the resulting displacement appears to be the superposition (the sum) of a wave traveling to the left and a wave traveling to the right. Use as the initial displacement the periodic curve illustrated in the figure. [*Hint:* Remember that we are concerned only with what happens in the interval $0 < x < L$ and $t > 0$.]

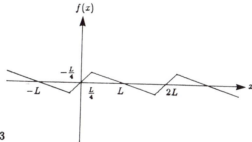

FIGURE FOR EXERCISE 3

4. A spherically symmetric wave has no angular dependence and must satisfy the three-dimensional wave equation

$$\frac{1}{r^2} \frac{\partial}{\partial r}\left(r^2 \frac{\partial u}{\partial r}\right) = \frac{1}{c^2} \frac{\partial^2 u}{\partial z^2}.$$

   Note that the normal angular terms which one would expect are missing. Transform this equation into a more familiar by assuming

$$u(r,\ t) = \frac{s(r,\ t)}{r}.$$

APPLICATION 14.2: Transistors and Characteristic Curves

See Section 14.6
Directional Derivatives and Gradients
Calculus, 4th Edition,
Larson/Hostetler/Edwards

A reasonable model for the characteristic curves for a model transistor can be obtained from the equation

$$I_c = 1.5 \ln[V_{ce}e^{I_b} + 1].$$

Characteristic curves show the relationship between the collector-emitter current I_c and voltage V_{ce}, and they show how these relate to the base current I_b. Actual curves are obtained experimentally. Figure 14.6 is such a set of curves. Currents here are assumed to be in milliamperes.

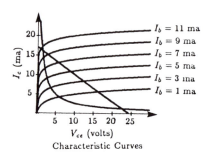

Characteristic Curves

FIGURE 14.6

Two other curves are commonly included on a graph of characteristic curves such as Figure 14.6. One is a hyperbola, representing the constant dissipation at a particular temperature. This represents the heat the unit can dissipate at room temperature without being destroyed by excessive heat. (In our example, $I_cV_{ce} = 25.5$ milliwatts.) The other curve shows the absolute maximum operating range with an infinite heat sink attached to the transistor so there is no build-up of heat. This is given by $I_c = -0.708V_{ce} + 17$.

For the protection of the transistor and to achieve the largest changes in output values, we would like to operate along the dissipation curve and have the values of I_c and V_{ce} move along the dissipation curve. This will happen when a particular characteristic curve is perpendicular to the dissipation curve.

EXAMPLE

Using Characteristic Curves

Under the above assumptions, at what point should the model transistor be operated if attached to a heat sink?

SOLUTION

To satisfy our conditions, we must have the directional derivative of an appropriate characteristic curve for a particular I_b to be perpendicular to the absolute maximum operating curve. This also requires knowledge of I_t and V_{ce}, which must simultaneously satisfy the perpendicular condition and intersection of the curves. The equations that must be solved are

$$1.5(0.708)e^{I_b} = V_{ce}e^{I_b} + 1$$

and

$$-0.708V_{ce} + 17 = 1.5\ln[V_{ce}e^{I_b} + 1].$$

Using an iterative procedure on a computer, we obtain

$$V_{ce} = 1.062, \qquad I_b = 10.77, \qquad \text{and } I_c = 16.25.$$

APPLICATION 14.2 EXERCISES

1. Find the equilibrium point for the 25° Celsius dissipation curve for the given example where the (constant) I_b curve is perpendicular to the dissipation curve.

2. Find the point for the 25° Celsius dissipation for the transistor illustrated in the figure. The equations associated with this model are

$$I_c = 1.5\ln[2V_{ce}e^{I_b} + 1]$$
$$I_cV_{ce} = 19.4$$

and

$$I_c = -0.85V_{ce} + 17.$$

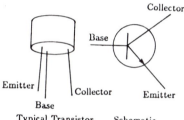

Emitter   Collector   Emitter

Base

Typical Transistor   Schematic

FIGURE

3. Find the ideal operating point for the model transistor of Exercise 2 when operating at the absolute maximum range. (See figure.)

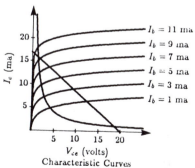

FIGURE FOR EXERCISES 2 AND 3

APPLICATION 14.3: Electric Field

> See Section 14.6
> **Directional Derivatives and Gradients**
> **Calculus, 4th Edition,**
> **Larson/Hostetler/Edwards**

Determining the direction in which charged particles move is important in a number of physical applications. One such case involves the cathode ray tube in a television set, as shown in Figure 14.7.

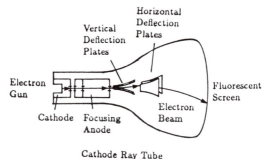

FIGURE 14.7

Cathode Ray Tube

EXAMPLE
━━━━━

Direction of Charged Particles

We consider here the simple case of a positively charged particle that is placed in an electric field produced by a dipole. The dipole consists of a pair of equal and opposite charges, $q = 1$ coulomb and $-q = -1$ coulomb, located 10 meters apart in the plane at $(5, 0)$ and $(-5, 0)$. If a positively charged particle is placed at $(5, 5)$, in which direction will it move? (See Figure 14.8.)

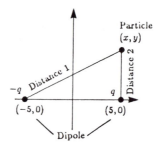

FIGURE 14.8

SOLUTION

We assume that a positively charged particle in an electric field always moves in a direction in which the voltage is decreasing most rapidly and that the voltage V at a point (x, y) is given by

$$V = \frac{q}{4\pi\epsilon_0}\left[\frac{1}{\sqrt{(x-5)^2 + y^2}} - \frac{1}{\sqrt{(x+5)^2 + y^2}}\right]$$

$$= \frac{q}{4\pi\epsilon_0}\left[\frac{1}{\text{distance 2}} - \frac{1}{\text{distance 1}}\right].$$

The voltage is decreasing most rapidly in the direction of $-\nabla V$. Since

$$V = \frac{q}{4\pi\epsilon_0}\left([(x-5)^2 + y^2]^{-1/2} - [(x+5)^2 + y^2]^{-1/2}\right)$$

we have

$$\nabla V = \frac{\partial V}{\partial x}\mathbf{i} + \frac{\partial V}{\partial y}\mathbf{j}$$

$$= \frac{q}{4\pi\epsilon_0}\left[\frac{-(x-5)}{[(x-5)^2 + y^2]^{3/2}} + \frac{x+5}{[(x+5)^2 + y^2]^{3/2}}\right]\mathbf{i}$$

$$+ \frac{q}{4\pi\epsilon_0}\left[\frac{-y}{[(x-5)^2 + y^2]^{3/2}} + \frac{y}{[(x+5)^2 + y^2]^{3/2}}\right]\mathbf{j}.$$

We evaluate at $(5, 5)$ to obtain

$$\nabla V = \frac{q}{4\pi\epsilon_0}\left[\frac{10}{(125)^{3/2}}\mathbf{i} + \left(-\frac{1}{25} + \frac{5}{(125)^{3/2}}\right)\mathbf{j}\right].$$

Thus, the direction that the particle will move is

$$-\frac{q}{4\pi\epsilon_0}\left[\frac{10}{(125)^{3/2}}\mathbf{i} + \left(-\frac{1}{25} + \frac{5}{(125)^{3/2}}\right)\mathbf{j}\right].$$

APPLICATION 14.3 EXERCISES

1. Consider the given example. If a positively charged particle is placed at the point $(-5, 5)$, in which direction will it move?
2. Consider the given example. If a positively charged particle is placed on the y-axis, show that the particle will move perpendicular to the y-axis.
3. Solve the given example except assume that the dipole charges are located at the points $(a, 0)$ and $(-a, 0)$ and the particle is located at (x, y). Assume the formula

$$V = \frac{q}{4\pi\epsilon_0}\left[\frac{1}{\sqrt{(x-a)^2 + y^2}} - \frac{1}{\sqrt{(x+a)^2 + y^2}}\right].$$

4. Consider the figure. The voltage in a vacuum between a long, straight, inner wire of radius a and a coaxial outer conductor that has inner radius b is given by

$$V = V_0\frac{\ln(r/a)}{\ln(b/a)}$$

where r is the distance from the common axis and V_0, $(V_0 > 0)$, is the potential of the outer cylinder. If a positively charged particle is placed at a distance r from the common axis where $a < r < b$, in which direction will it move?

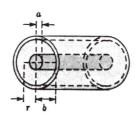

FIGURE FOR EXERCISE 4

APPLICATION 14.4: Heat Loss for Buildings

> **See Section 14.9**
> **Applications of Extrema of Functions of Two Variables**
> **Calculus, 4th Edition,**
> **Larson/Hostetler/Edwards**

Energy-conscious people are aware that poorly insulated or uninsulated buildings can lose a considerable amount of heat, thereby wasting both money and natural resources. But there may be other motives for minimizing heat loss. Computers and other sensitive equipment require a stable temperature environment. In such cases, minimizing heat loss is essential to proper functioning.

Various materials and temperature controlling devices can be used to help curtail heat loss; the architecture of the room or building itself can also play a role.

EXAMPLE

Minimizing Heat Loss

A heated storage room in International Falls, Minnesota, contains sensitive equipment. It is shaped like a rectangular box and has a volume of 1000 cubic feet. (See Figure 14.9.) Because warm air rises, in a given period of time the heat loss per unit area through the ceiling is five times as great as the heat loss through the floor. If the heat loss through the walls is three times as great as the heat loss through the floor, determine the room dimensions that will minimize heat loss and thus minimize heating costs.

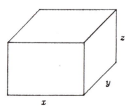

FIGURE 14.9

SOLUTION

Let x represent the length, y the width, and z the height of the room. Let V represent the volume of the room. We begin with the equation $V = xyz = 1000$. We calculate the heat loss H through each of the six sides to get

$$H = (5xy + xy + 3xz + 3xz + 3yz + 3yz)k$$
$$= (6xy + 6xz + 6yz)k$$

where k is a constant representing the heat loss per unit area in a given period of time for the floor. We eliminate a variable by substituting $z = 1000/xy$ for H to obtain

$$H = \left(6xy + \frac{6000}{y} + \frac{6000}{x}\right)k$$

where $0 < x < \infty$ and $0 < y < \infty$. Next, to minimize H, we evaluate partial derivatives and set them to zero.

$$\frac{\partial H}{\partial x} = \left(6y - \frac{6000}{x^2}\right)k = 0$$

$$\frac{\partial H}{\partial y} = \left(6x - \frac{6000}{y^2}\right)k = 0$$

We solve simultaneously. From the first equation we have $y = 1000/x^2$ and so,

$$x - 1000\left(\frac{x^4}{1000^2}\right) = 0$$

$$x(1000 - x^3) = 0$$

$$x = 0 \quad \text{or} \quad x = 10.$$

Hence, $y = 1000/100 = 10$ and $z = 1000/(10)(10) = 10$. We must show that this result represents a minimum.

$$\frac{\partial^2 H}{\partial x^2} = \frac{12000}{x^3} = \frac{12000}{1000} = 12 > 0$$

$$\frac{\partial^2 H}{\partial y^2} = \frac{12000}{y^3} = \frac{12000}{1000} = 12 > 0$$

$$\frac{\partial^2 H}{\partial x \partial y} = 6 > 0$$

Thus, the discriminant

$$D = (H_{xx})(H_{yy}) - (H_{xy}) = 144 - 36 = 108 > 0.$$

Since $H_{xx} > 0$, we know that $x = 10$, $y = 10$, and $z = 10$ represents a minimum. Thus, a storeroom that is cubical in shape will minimize heat loss for the room under discussion. Note that the constant k cancelled out and thus had no effect on the answer. Only relative heat loss was important.

APPLICATION 14.4 EXERCISES

1. Redo the example using Lagrange multipliers.
2. Redo the example assuming that there is no heat loss through the floor.
3. Insulation can be added to the ceiling to reduce the heat loss through the ceiling. Assume that the heat flow through the ceiling in the example can be made arbitrarily low through the addition of more insulation. Is it possible to have the minimum heat loss occur when the height z is 8 feet?
4. Redo the example assuming that the heat loss through one wall (the northern wall) is four times as great as the heat loss through the floor.

APPLICATION 14.5: Hotplates

> See Section 14.9
> **Applications of Extrema of Functions of Two Variables**
> **Calculus, 4th Edition,**
> **Larson/Hostetler/Edwards**

The temperature of a physical object, like a hotplate, depends in part on the physical shape of the object and on the temperature distribution function T. We can use calculus to find the location for maximum and minimum temperatures when in equilibrium.

Suppose T is a harmonic function. (This can be verified by observing that the Laplacian $T_{xx} + T_{yy} = 0$.) We can express the temperature distribution on the boundary in polar coordinates using the equation for the shape of the object (hotplate), also expressed in polar coordinates. The resulting equation can be analyzed for critical points and the maximum and minimum values determined.

EXAMPLE

Hotplate

A circular disk is placed between two fires and comes into thermodynamic equilibrium. The equilibrium temperature is given by

$$T(x, y) = x^2 - y^2 - x$$

where the disk is defined by $x^2 + y^2 = 25$. Where is the temperature on the disk the lowest?

SOLUTION

Our function T is harmonic. This can be verified by calculating the Laplacian of T.

$$T_x = 2x - 1$$
$$T_{xx} = 2$$

In a similar way, we get

$$T_y = -2y$$
$$T_{yy} = -2.$$

A simple comparison shows that the Laplacian $T_{xx} + T_{yy} = 0$. Since the function is harmonic, the maximum is on the boundary. Rewriting the temperature distribution on the boundary using polar coordinates, we get

$$T(r, \theta) = (r \cos \theta)^2 - (r \sin \theta)^2 - r \cos \theta.$$

Minimizing this expression is equivalent to finding the minimum value for T as a function of the single variable θ.

$$T(5, \theta) = 25 \cos^2 \theta - 25 \sin^2 \theta - 5 \cos \theta$$
$$= t(\theta)$$

We can obtain the critical points of t by standard one-dimensional calculus techniques.

$$t'(\theta) = -100 \sin \theta \left(\cos \theta - \frac{1}{20} \right)$$

The critical points are therefore at $\theta = 0$, π, and $\cos^{-1} \frac{1}{20}$. The resulting values for t are 20, 30, and -25.125, respectively.

APPLICATION 14.5 EXERCISES

1. Find the location of the minimum and maximum temperatures for the temperature distribution in the example, $T(x, y) = x^2 - y^2 - x$, for each of the four objects illustrated in the figure.

2. Find the location of the minimum and maximum temperature for each of the four objects illustrated in the figure for the temperature distribution $T(x, y) = x^3 - 3xy^2$.

3. Find the location of the minimum and maximum temperature for each of the objects illustrated in the figure for the temperature distribution $T(x, y) = x^4 - 6x^2y^2 + y^4$.

4. Given

$$U(x, y) = \frac{x + y}{x^2 + y^2}$$

explain why the method shown in the example and solution is not acceptable for the regions shown in the figure even though U appears to satisfy Laplace's equation.

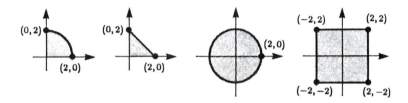

FIGURE FOR EXERCISES 1–4

15

MULTIPLE INTEGRATION

APPLICATION 15.1: Sprinkler

> **See Section 15.3**
> **Change of Variables: Polar Coordinates**
> **Calculus, 4th Edition,**
> **Larson/Hostetler/Edwards**

During the drought of 1989, many homeowners were forced to cut back the amount of time that their lawn sprinklers were in use. Several homeowners discovered that some parts of their lawns remained green while other parts became brown. The problem was that sprinklers do not put out a uniform amount of water. Suppose the homeowner's sprinkler distributes water in a radial fashion according to the formula

$$f(r) = \frac{r}{16} - \frac{r^2}{160}$$

cubic feet of water per hour per square foot of lawn where r is the distance in feet from the sprinkler. For this sprinkler, the distribution of water can be shown to be different for two regions of the lawn.

EXAMPLE

Lawn Sprinkler

Consider a circular lawn with a radius of 10 feet, as shown in Figure 15.1. Assume that the sprinkler is at the center of the lawn. Find the amount of water that is distributed in one hour in the two following annular regions.

$$A = \{(r, \theta) : 4 \le r \le 5, \ 0 \le \theta \le 2\pi\}$$
$$B = \{(r, \theta) : 9 \le r \le 10, \ 0 \le \theta \le 2\pi\}$$

Is the distribution of the water uniform?

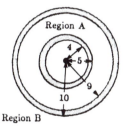 

FIGURE 15.1

SOLUTION

The amount of water deposited onto region A of the lawn in one hour is given by

$$V_A = \int_0^{2\pi} \int_4^5 \left[\frac{r}{16} - \frac{r^2}{160} \right] r \, dr \, d\theta$$

$$= 2\pi \left[\frac{r^3}{48} - \frac{r^4}{640} \right]_4^5$$

$$\approx 4.36 \text{ cubic feet.}$$

Similarly, the amount of water deposited in one hour onto region B of the lawn is $V_B \approx 1.71$ cubic feet. Since the area of region A is $A_A = 9\pi$ and the area of region B is $A_B = 19\pi$, we see that in one hour region A receives

$$\frac{V_A}{A_A} = \frac{4.36}{9\pi} \approx 0.154$$

cubic feet of water per square foot of lawn, whereas region B receives

$$\frac{V_B}{A_B} = \frac{1.71}{19\pi} \approx 0.029$$

cubic feet of water per square foot of lawn. Thus, the distribution of water in these two regions is not uniform.

APPLICATION 15.1 EXERCISES

1. Determine the amount of water that the entire lawn receives in one hour.

2. Because of the drought, the homeowner in the example is restricted to 600 cubic feet of outside watering for each nine-week billing period. How many hours per week can the homeowner use the lawn sprinkler?

3. Which annular region $\{(r, \theta) : a \leq r \leq a+1,\ 0 \leq \theta \leq 2\pi\}$, where $0 \leq a \leq 9$ receives the most water per square foot?

4. Which annular region in Exercise 3 receives the least water per square foot?

APPLICATION 15.2: Inertia and Center of Mass

> **See Section 15.4**
> **Center of Mass and Moments of Inertia**
> **Calculus, 4th Edition,**
> **Larson/Hostetler/Edwards**

A car skids on ice and spins around. An air hockey game disk spins as it glides across the table. The earth and moon rotate as a two-body system. Each of these situations shows a rotation about some fixed point, but what would that point be?

EXAMPLE

Point of Rotation

Consider a frictionless plane on which an object is moving. Such an object, free to rotate, would be expected to do so about some fixed point. Assume the object is viewed as two-dimensional, with variable density. By considering the minimum energy for the system, calculate the location of the point about which the object will rotate with a fixed angular velocity ω.

SOLUTION

The energy of a rotating body, without translation, is given by

$$E = \frac{1}{2} I \omega^2.$$

For our two-dimensional object with variable density ρ, the moment of inertia about any point $(x_0,\ y_0)$ perpendicular to the surface on which the object is moving is given by

$$I = \iint_R [(x - x_0)^2 + (y - y_0)^2] \rho(x,\ y)\, dA.$$

Since we want the minimum energy, and assuming the angular velocity is constant, it suffices to find the minimum moment of inertia. To obtain this result, we take partial derivatives with respect to x_0 and y_0 and then look for the critical points. Setting these partial derivatives equal to zero produces

$$\frac{\partial I}{\partial x_0} = -\iint_R 2(x - x_0) \rho(x,\ y)\, dA = 0$$

and

$$\frac{\partial I}{\partial y_0} = -\iint_R 2(y - y_0) \rho(x,\ y)\, dA = 0.$$

These equations can be rearranged as

$$\iint_R x\rho(x,\ y)\, dA = \iint_R x_0 \rho(x,\ y)\, dA = x_0 \iint_R \rho(x,\ y)\, dA = x_0 m$$

and

$$\iint_R y\rho(x,\ y)\, dA = \iint_R y_0 \rho(x,\ y)\, dA = y_0 \iint_R \rho(x,\ y)\, dA = y_0 m.$$

Rewriting once again while solving for x_0 and y_0, we get

$$x_0 = \frac{\displaystyle\iint_R x\rho(x,\ y)\, dA}{m} = x\text{-coordinate of the center of mass}$$

and

$$y_0 = \frac{\displaystyle\iint_R y\rho(x,\ y)\, dA}{m} = y\text{-coordinate of the center of mass.}$$

It is not difficult to verify that the solution is indeed a minimum. We need only apply the second derivative test for functions of two variables to our results. The second partial derivatives are

$$\frac{\partial^2 I}{\partial x_0{}^2} = 2\iint_R \rho(x,\ y)\, dA = 2m$$

$$\frac{\partial^2 I}{\partial y_0{}^2} = 2\iint_R \rho(x,\ y)\, dA = 2m$$

$$\frac{\partial^2 I}{\partial x_0 \partial y_0} = 0$$

$$\frac{\partial^2 I}{\partial y_0 \partial x_0} = 0$$

where m is the mass. We next consider the determinant

$$d = \begin{vmatrix} \dfrac{\partial^2 I}{\partial x_0{}^2} & \dfrac{\partial^2 I}{\partial x_0 \partial y_0} \\[2ex] \dfrac{\partial^2 I}{\partial y_0 \partial x_0} & \dfrac{\partial^2 I}{\partial y_0{}^2} \end{vmatrix}$$

which gives us

$$4\left[\iint_R \rho(x,\ y)\, dA\right]^2 = 4m^2 > 0.$$

Since the partial derivative, $\partial^2 I/\partial x_0{}^2 > 0$, we have a minimum. The results here can easily be applied to other situations. For example, considering the Earth and the Moon as a two-body problem (ignoring the Sun, Jupiter, and all other heavenly bodies), the point about which the system will rotate is the center of mass of the two objects.

APPLICATION 15.2 EXERCISES

1. The Moon and the Earth appear to rotate about a common point. Find the location of this point by minimizing the moment of inertia of the Moon-Earth system. The mass of the Moon is 7.36×10^{22} kilograms and the mass of the Earth is 5.98×10^{24} kilograms. The distance between the Earth and the Moon is 3.82×10^{8} meters.

2. Two ice skaters are holding hands and revolving together, arms extended. The man's mass is 80 kilograms, the woman's is 50 kilograms, and the distance between them is 1.8 meters. Treat the skaters as point masses. Minimize the energy for the rotating system to find the center of mass for the skating pair.

3. A baton is a long rod with a weighted sphere on one end. Assume the rod is 1 meter long, is of uniform density, and has a total mass of 0.5 kilograms. The weighted sphere is 0.2 kilograms. Find the center of mass for the baton by minimizing the moment of inertia.

APPLICATION 15.3: Moment of Inertia and the Pendulum

> See Section 15.4
> Center of Mass and Moments of Inertia
> Calculus, 4th Edition,
> Larson/Hostetler/Edwards

In what way is a baseball bat like a pendulum? While they both "swing" and have "pivot points," there is another, less obvious correspondence. When a batter connects with the ball at a specific point m on the bat, called the center of percussion, the batter's hands do not sting. The specific location L of that center of percussion can be found by analyzing a pendulum.

EXAMPLE

Pendulum Center of Oscillation

The equation of motion for the pendulum can be obtained with Newton's Equations of Motion. Either the force or torque equation may be used. For the physical pendulum, the torque equation is more appropriate. A simple pendulum replacement can be made for a physical pendulum that will result in the same period. Consider l to be the replacement length for the simple pendulum system. What is this location of L, the center of oscillation?

SOLUTION

The physical pendulum is shown in Figure 15.2.

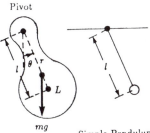

FIGURE 15.2

To begin we write

$$\tau = I\frac{d^2\theta}{dt^2}$$

where τ is the torque, θ is the angular displacement from the vertical, t is the time, and I is the moment of inertia about the pivot. The torque that tries to restore the system in Figure 15.2 is $\mathbf{r} \times \mathbf{F}$ where $\mathbf{r}$ is the radial vector from the pivot and $\mathbf{F}$ is the force due to gravity acting on the center of mass. The Parallel Axis Theorem states that the moment of inertia I through some axis a distance r from the center of mass is given by

$$I = mr^2 + I_{cm}$$

where I_{cm} is the moment of inertia through the center of mass and the two axes are parallel. Thus, we can write

$$-rmg\sin\theta = \left(mr^2 + I_{cm}\right)\frac{d^2\theta}{dt^2}$$

which implies (for small values of θ) that

$$\left(mr^2 + I_{cm}\right)\frac{d^2\theta}{dt^2} + mgr\theta \approx 0,$$

where r is the distance between the pivot and the center of mass, m is the mass, g is the acceleration due to gravity, and θ is the angle. The solution of this "equation" is

$$\theta(t) = A\cos\left(\sqrt{\frac{mgr}{mr^2 + I_{cm}}}\,t + \phi\right)$$

where ϕ and A are constants. This implies that the frequency of the pendulum is

$$f = \frac{1}{2\pi}\sqrt{\frac{mgr}{mr^2 + I_{cm}}}.$$

The distance from the pivot to the place where a simple bob would oscillate at the same frequency is given by

$$\frac{1}{2\pi}\sqrt{\frac{mgr}{mr^2 + I_{cm}}} = \frac{1}{2\pi}\sqrt{\frac{g}{l}}.$$

Solving for l yields

$$l = \frac{mr^2 + I_{cm}}{mr}.$$

The point L is called the center of oscillation because it is the replacement location (using the original pivot) for a simple pendulum with the same frequency. If the physical pendulum were to pivot about the point L instead of the original pivot, it would oscillate with the same frequency. This point also corresponds to the center of percussion. When the ball hits the bat at the center of percussion, the batter's hands do not sting.

APPLICATION 15.3 EXERCISES

1. Derive the Parallel Axis Theorem which states that the moment of inertia I through any point is the same as the moment of inertia through the center of mass I_{cm} plus the mass times the square of the distance the axis is shifted. That is, $I = I_{cm} + mr^2$. As the theorem states, the new axis of rotation is parallel to the one through the center of mass.

2. The figure shows a typical baseball bat. Assuming that the batter holds the bat four inches from the end, where should the ball hit so there is no sting?

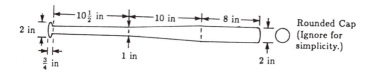

FIGURE FOR EXERCISE 2

3. As an experiment, get a baseball bat, hold it in one hand and tap the large end against your shoe heel or some other object that won't be damaged. See if you can locate the center of percussion. Does the location you find agree with the calculated location?

4. In tennis, people talk about the "sweet spot" of a tennis racket. The sweet spot is another name used for the center of percussion. Assume that your hand is centered 2 inches from the handle end. Ignore the mass of the strings and assume that the material used to make the racket is of uniform density. Find the "sweet spot" of the racket in figure.

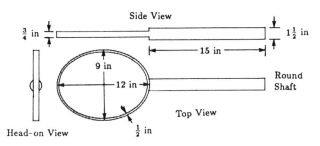

FIGURE FOR EXERCISE 4

APPLICATION 15.4: Surface Area

> See Section 15.5
> **Surface Area**
> **Calculus, 4th Edition,**
> **Larson/Hostetler/Edwards**

There is a major department store in Foster, California, at the Fashion Island shopping center. The roof of part of the building is considered unusual by most. It looks like a tent with numerous poles supporting it. We want to model a tent-like roof. Our eventual objective is to calculate the surface area that would be necessary for resurfacing such a roof.

EXAMPLE

Surface Area of an Unusual Roof

Assume that the roof has evenly spaced peaks in two directions, and no valleys to collect water during storms. There are horizontal flat regions in the construction but they drain the water that accumulates. (See Figure 15.3.)

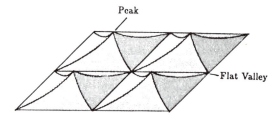

Shopping Center Roof

FIGURE 15.3

SOLUTION

We begin our analysis by sketching a reasonable single cell of such a roof (Figure 15.4). The entire roof is made of several such cells connected to each other.

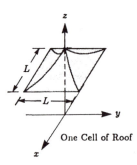

One Cell of Roof

FIGURE 15.4

The function we use for our model is the hyperbolic cosine. We choose this function because it describes the hanging cable, a configuration that provides a uniform horizontal load, and less stress for the one structure. In our model, the height z will be given by

$$z = h \cosh \left[a^2 \left(\frac{L}{2} - |x| \right) \left(\frac{L}{2} - |y| \right) \right]$$

with $-L/2 \leq x \leq L/2$ and $-L/2 \leq y \leq L/2$. The difference in height between the edges and the center is

$$h \cosh\left(\frac{a^2 L^2}{4}\right).$$

Using this model, find the surface of the roof.

SOLUTION

By symmetry, we need only consider one-fourth of the basic cell, $0 \leq x \leq L/2$ and $0 \leq y \leq L/2$. The resulting integral is then

$$S = 4 \int_0^{L/2} \int_0^{L/2} \sqrt{1 + z_x{}^2 + z_y{}^2} \, dx \, dy$$

$$= 4 \int_0^{L/2} \int_0^{L/2} \sqrt{1 + h^2 a^4 \left[\left(\frac{L}{2} - x\right)^2 + \left(\frac{L}{2} - y\right)^2\right] \sinh^2\left[a^2\left(\frac{L}{2} - x\right)\left(\frac{L}{2} - y\right)\right]} \, dx \, dy.$$

Now, let $t = a[(L/2) - y]$ and $s = a[(L/2) - x]$. Then the surface integral yields

$$S = 4 \int_0^{aL/2} \int_0^{aL/2} \frac{1}{a^2} \sqrt{1 + h^2 a^2 (t^2 + s^2) \sinh^2(st)} \, ds \, dt$$

which will give the surface area of one entire cell.

APPLICATION 15.4 EXERCISES

1. Evaluate the integral given in the example. Let $L = 100$, $h = 25$, and $a = \frac{1}{50}$. What is the total surface area of the text? [Hint: Use Simpson's Rule as an iterated integral to evaluate the surface area of $\frac{1}{4}$ of a roof cell. Divide the region into 16 subdivisions in order to use Simpson's Rule with $n = 4$ in each direction.]

2. Calculate the amount of material needed to make the "dome" tent illustrated in the figure. Allow an additional $\frac{1}{2}$-inch on each edge for seams. The tent also has a floor of the same material.

6-sided Dome Tent

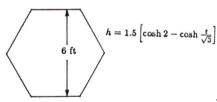

Base (regular hexagon)

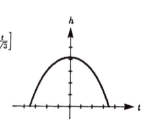

$$h = 1.5\left[\cosh 2 - \cosh \frac{t}{\sqrt{3}}\right]$$

End View of Tent

FIGURE FOR EXERCISE 2

3. A small observatory has half of an ellipse of revolution for the roof over its telescope. The roof has a height of 10 meters and a diameter of 16 meters. Find the number of square meters of aluminum needed to make the skin of the roof.

APPLICATION 15.5: Gravitational Force Revisited

> See Section 15.7
> Triple Integrals in Cylindrical and Spherical Coordinates
> Calculus, 4th Edition,
> Larson/Hostetler/Edwards

Newton's Law of Gravitation describes the force between two objects of mass m and mass M. Frequently, we model a physical situation by replacing a physical mass by a point mass located at the center of mass of the original object. For example, in computing the fluid force on a submerged object, one uses the average depth of the object in place of the actual depths. We consider an example where this principle cannot be applied.

EXAMPLE

Gravitational Force Between Cylinder and Particle

An astronaut has accidentally left a cylindrical can in space. A particle of dust is nearby. We assume Newton's Law of Gravitation

$$\|\mathbf{F}\| = \frac{GmM}{d^2}$$

where G is the universal gravitational constant, and d is the distance between the centers of mass of the two masses m and M. Let the cylinder (with constant density ρ) be given by

$$\{(x, \ y, \ z) : x^2 + y^2 \leq a^2, \ b \leq z \leq b + h, \ b > 0\}.$$

Note that the cylinder has a radius of a and a height of h. Assume that the particle of dust is located at the origin. (See Figure 15.5.) Describe the gravitational force between the particle and can.

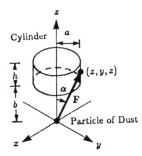

FIGURE 15.5

SOLUTION

By symmetry, the x and y components of the gravitational force are 0. We need only find the z component of the force. This component is

$$\Delta F_z \approx \frac{Gm\rho \cos \alpha}{d^2} \, \Delta V$$

where α is the angle between the positive z-axis and the radial vector at the point $(x, \ y, \ z)$. Using $\cos \alpha = z/d$ and converting to cylindrical coordinates, we have

$$\Delta F_z \approx \frac{Gm\rho z}{(r^2 + z^2)^{3/2}} r \Delta r \Delta \theta \Delta z.$$

Thus,

$$F_z = Gm\rho \int_b^{b+h} \int_0^{2\pi} \int_0^a \frac{rz}{(r^2 + z^2)^{3/2}} \, dr \, d\theta \, dz$$

$$= Gm\rho \int_b^{b+h} \int_0^{2\pi} \left[-\frac{z}{(r^2 + z^2)^{1/2}} \right]_0^a \, d\theta \, dz$$

$$= Gm\rho \int_b^{b+h} \int_0^{2\pi} \left[1 - \frac{z}{(a^2 + z^2)^{1/2}} \right] \, d\theta \, dz$$

$$= 2\pi Gm\rho \int_b^{b+h} \left[1 - \frac{z}{(a^2 + z^2)^{1/2}} \right] \, dz$$

$$= 2\pi Gm\rho \left[z - (a^2 + z^2)^{1/2} \right]_b^{b+h}$$

$$= 2\pi Gm\rho \left(h - \sqrt{a^2 + (b+h)^2} + \sqrt{a^2 + b^2} \right).$$

Replacing the cylinder by a point mass located at its geometric center would yield

$$F_z = \pi a^2 h \frac{\rho G m}{[b + (h/2)]^2}.$$

(See Exercise 1.) Clearly these answers do not agree. This is an example where replacing an object by a point mass is not valid. However, if the cylinder is very far from the point mass (b is very large relative to a and h), then the forces are nearly equal. (See Exercise 2.)

APPLICATION 15.5 EXERCISES

1. Assume that the cylinder is condensed to a point mass located at its geometric center. Calculate the gravitational force exerted by this condensed point mass on the original mass m.

2. Assume that b is very large (relative to a and h). Use the Binomial Theorem to approximate the force on a point mass m due to the cylinder.

3. Work the example but use the cylindrical shell $\{(x, y, z) : a^2 \le x^2 + y^2 \le b^2, \ 0 \le z \le h\}$.

4. Work the example but use the cylinder $\{(x, y, z) : 0 \le x^2 + y^2 \le a^2, \ 0 \le z \le h\}$. (You will need a limiting argument.)

16

VECTOR ANALYSIS

APPLICATION 16.1: Heat Flux Across a Boundary

> See Section 16.5
> Surface Integrals
> Calculus, 4th Edition,
> Larson/Hostetler/Edwards

We wish to measure the heat flux (flow) across the surface of an object. Heat flows from higher to lower temperatures in the direction of greatest change. As a result, measuring the heat flux involves the gradient of the temperature. The flux depends on the area of the surface. It is the normal direction to this surface that is important because heat that flows in directions tangential to the surface will give no heat loss. Thus, we shall assume that the heat flux across a portion of the surface of area ΔS is given by

$$\Delta H \approx -k\nabla T \cdot \mathbf{n}\, \Delta S$$

where T is the temperature, $\mathbf{n}$ is the unit normal to the surface in the direction of the heat flow, and k is the thermal diffusivity of the material. Hence, the heat flux across the surface S is then given by

$$H = \iint_S -k\nabla T \cdot \mathbf{n}\, dS.$$

EXAMPLE

Heat Flux Across a Cylindrical Roof

Consider the case of a single heat source located at the origin with temperature

$$T(x,\ y,\ z) = \frac{25}{(x^2 + y^2 + z^2)^{1/2}}.$$

Calculate the heat flux across the surface

$$S = \left\{ (x,\ y,\ z) : z = \sqrt{1 - x^2},\ -\frac{1}{2} \le x \le \frac{1}{2}, 0 \le y \le 1 \right\}.$$

(See Figure 16.1.)

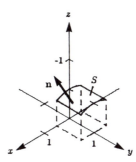

FIGURE 16.1

SOLUTION

We begin by determining ∇T, $\mathbf{n}$, and dS.

$$\nabla T = \frac{\partial T}{\partial x}\mathbf{i} + \frac{\partial T}{\partial y}\mathbf{j} + \frac{\partial T}{\partial z}\mathbf{k}$$

$$= 25\left[-\frac{x}{(x^2 + y^2 + z^2)^{3/2}}\mathbf{i} - \frac{y}{(x^2 + y^2 + z^2)^{3/2}}\mathbf{j} - \frac{z}{(x^2 + y^2 + z^2)^{3/2}}\mathbf{k}\right]$$

$$\mathbf{n} = \frac{-\dfrac{\partial z}{\partial x}\mathbf{i} - \dfrac{\partial z}{\partial y}\mathbf{j} + \mathbf{k}}{\sqrt{\left(\dfrac{\partial z}{\partial x}\right)^2 + \left(\dfrac{\partial z}{\partial y}\right)^2 + 1}} = \frac{\dfrac{x\mathbf{i}}{\sqrt{1 - x^2}} + \mathbf{k}}{\sqrt{\dfrac{x^2}{1 - x^2} + 1}} = x\mathbf{i} + \sqrt{1 - x^2}\,\mathbf{k}$$

$$dS = \sqrt{\left(\frac{\partial z}{\partial x}\right)^2 + \left(\frac{\partial z}{\partial y}\right)^2 + 1}\,dy\,dx = \sqrt{\frac{x^2}{1 - x^2} + 1}\,dy\,dx = \frac{1}{\sqrt{1 - x^2}}\,dy\,dx$$

Thus, the heat flux across the surface S is

$$H = \iint_S -k\nabla T \cdot \mathbf{n}\,dS$$

$$= 25k\iint_R \left[\frac{x^2}{(x^2 + y^2 + z^2)^{3/2}(1 - x^2)^{1/2}} + \frac{z}{(x^2 + y^2 + z^2)^{3/2}}\right]dA$$

$$= 25k\int_{-1/2}^{1/2}\int_0^1 \left[\frac{x^2}{(1 + y^2)^{3/2}(1 - x^2)^{1/2}} + \frac{1 - x^2}{(1 + y^2)^{3/2}(1 - x^2)^{1/2}}\right]dy\,dx$$

$$= 25k\int_{-1/2}^{1/2}\int_0^1 \frac{1}{(1 + y^2)^{3/2}(1 - x^2)^{1/2}}\,dy\,dx$$

$$= 25k\int_0^1 \frac{1}{(1 + y^2)^{3/2}}\,dy\int_{-1/2}^{1/2}\frac{1}{(1 - x^2)^{1/2}}\,dx.$$

The region R is shown in Figure 16.2.

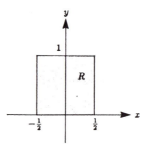

FIGURE 16.2

Since

$$\int_0^1 \frac{dy}{(1+y^2)^{3/2}} = \int_0^{\pi/4} \frac{\sec^2 \theta}{\sec^3 \theta} \, d\theta$$

$$= \int_0^{\pi/4} \frac{1}{\sec \theta} \, d\theta$$

$$= \int_0^{\pi/4} \cos \theta \, d\theta$$

$$= \sin \theta \bigg]_0^{\pi/4} = \frac{\sqrt{2}}{2}$$

and

$$\int_{-1/2}^{1/2} \frac{1}{(1-x^2)^{1/2}} \, dx = \sin^{-1} x \bigg]_{-1/2}^{1/2} = \frac{\pi}{3}$$

we have

$$H = 25k \left(\frac{\pi}{3}\right) \left(\frac{\sqrt{2}}{2}\right).$$

If the surface S represents a glass skylight and the dimensions of S are in meters, then $k = 4 \times 10^{-4}$ calories per second $\left(\text{cm}^2 \text{ degrees Celsius}\right)$ and $H \approx 74$ calories per second, where T is measured in degrees Celsius.

APPLICATION 16.1 EXERCISES

1. We can parametrize the surface in this example by letting $x = \cos u$, $y = v$, $z = \sin u$, where $\pi/3 \leq u \leq 2\pi/3$, and $0 \leq v \leq 1$. Then, if $\mathbf{r} = x\mathbf{i} + y\mathbf{j} + z\mathbf{k}$,

$$\pm\mathbf{n} = \frac{\dfrac{\partial \mathbf{r}}{\partial u} \times \dfrac{\partial \mathbf{r}}{\partial v}}{\left\| \dfrac{\partial \mathbf{r}}{\partial u} \times \dfrac{\partial \mathbf{r}}{\partial v} \right\|} \quad \text{and} \quad dS = \left\| \dfrac{\partial \mathbf{r}}{\partial u} \times \dfrac{\partial \mathbf{r}}{\partial v} \right\| \, du \, dv.$$

Solve this problem by writing the integral in terms of u and v. (This is equivalent to using cylindrical coordinates with radius 1.)

2. Solve the given example for the hemisphere

$$S = \{(x, y, z) : z = \sqrt{1 - x^2 - y^2}, \ x^2 + y^2 \leq 1\}.$$

(See the figure.)

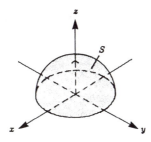

FIGURE FOR EXERCISE 2

3. Rework Exercise 2 using the parametrization $x = \sin u \cos v$, $y = \sin u \sin v$, $z = \cos u$, $0 \le u \le \pi/2$, and $0 \le v \le 2\pi$. (Note that this is equivalent to using spherical coordinates with radius 1.)

4. Solve the given example for the cylinder

$$S = \{(x, y, z) : x^2 + y^2 = 1, \ 0 \le z \le 1\}.$$

(See the figure.)

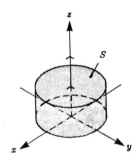

FIGURE FOR EXERCISE 4

APPLICATION 16.2: Divergence and Gauss's Law

> **See Section 16.6**
> **Divergence Theorem**
> **Calculus, 4th Edition,**
> **Larson/Hostetler/Edwards**

Suppose we have a surface with an enclosed charge, Q. Gauss's Law in integral form for electrostatics is

$$\iint_S \mathbf{E} \cdot \mathbf{n} \, dS = \frac{Q}{\epsilon_0}$$

where Q is the enclosed charge, $\mathbf{E}$ is the electric field, $\mathbf{n}$ is the outward unit normal, and ϵ_0 is a constant called the permittivity of free space. The integral over the surface that surrounds the enclosed charge is called the flux integral because it measures the amount of electric field that flows out of the enclosed region. We would like to be able to calculate the flux through the side of a cube with a charge Q at its center. We might employ the Divergence Theorem but unfortunately, the Divergence Theorem requires the partial derivatives for $\mathbf{E}$ to be continuous over the entire region enclosed by the surface. If Q is located at the origin, then the electric field from Q is given by Coulomb's Law

$$\mathbf{E} = \frac{Q}{4\pi\epsilon_0 r^2} \mathbf{r}$$

where $\mathbf{r}$ is a unit vector pointing away from the positive charge. This expression does not have continuous partial derivatives at the origin. On the other hand, because of Coulomb's Law, the integral

$$\iint_S \mathbf{E} \cdot \mathbf{n} \, dS$$

for the electric field due to a point charge at the origin is simple. We now have an approach to our problem.

EXAMPLE

Flux Calculation

Given a cube with point charge at its center, (see Figure 16.3) what is the flux through one side of the cube?

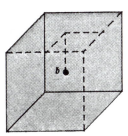

FIGURE 16.3

SOLUTION

We begin with the surface S of a sphere which contains a point charge Q at its center. Then

$$\mathbf{E} \cdot \mathbf{n} = E\mathbf{r} \cdot \mathbf{n} = E = \|\mathbf{E}\|.$$

So,

$$\iint_S \mathbf{E} \cdot \mathbf{n} \, dS = E \iint_S dS = 4\pi r^2 E$$

and

$$\iint_S \frac{Q}{4\pi\epsilon_0 r^2} \mathbf{r} \cdot \mathbf{r} \, dS = \frac{Q}{4\pi\epsilon_0} \iint_S \frac{r^2 \, d\Omega}{r^2} = \frac{4\pi Q}{4\pi\epsilon_0} = \frac{Q}{\epsilon_0}.$$

The integration over $d\Omega$, an element of solid angle, results in 4π. We still haven't used the Divergence Theorem nor, at this point, are we any closer to obtaining the flux through the side of a cube. We know that the divergence of $\mathbf{E}$ anywhere *excluding* the origin is zero. This can be shown to be true by first writing the divergence in spherical coordinates as

$$\nabla \cdot \mathbf{A} = \frac{1}{r^2} \frac{\partial}{\partial r}[r^2 A_r] + \frac{1}{r \sin \phi} \frac{\partial}{\partial \phi}[\sin \phi A_\phi] + \frac{1}{r \sin \phi} \frac{\partial A_\theta}{\partial \theta}$$

where A_r is the r component of A, A_ϕ is the ϕ component, and A_θ is the θ component. Since E is only radial, then $E = E(r)$ and

$$\nabla \cdot \mathbf{E} = \frac{1}{r^2} \frac{\partial}{\partial r}\left[r^2 \frac{Q}{4\pi\epsilon_0 r^2}\right] = 0.$$

Thus, by the Divergence Theorem, $\iint\limits_S \mathbf{E} \cdot \mathbf{n} \, dS = 0$ about any closed region excluding the origin. We also know that the integral $\iint\limits_S \mathbf{E} \cdot \mathbf{n} \, dS$, when the origin is included, gives Q/ϵ_0 so we next consider the special-shaped region in Figure 16.4.

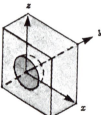

FIGURE 16.4

This region is half of a cube with a hemisphere cut out. The origin is centered on the side that has the hemisphere cut out. We know the flux through an entire spherical shell due to a point charge is Q/ϵ_0, so by symmetry, we know the flux through the hemispherical shell is $Q/2\epsilon_0$. We can integrate over the illustrated surface giving

$$\iint_S \mathbf{E} \cdot \mathbf{n} \, dS = 0$$

$$= \underbrace{\iint_S \mathbf{E} \cdot \mathbf{n} \, dS}_{\text{across hemisphere}} + 4\underbrace{\iint_S \mathbf{E} \cdot \mathbf{n} \, dS}_{\text{across half-sides}} + \underbrace{\iint_S \mathbf{E} \cdot \mathbf{n} \, dS}_{\text{across full side}} + \underbrace{\iint_S \mathbf{E} \cdot \mathbf{n} \, dS}_{\substack{\text{across plane with} \\ \text{circle cut out}}}.$$

We know that $\iint\limits_S \mathbf{E} \cdot \mathbf{n} \, dS$ for the hemispherical part is $-Q/2\epsilon_0$ and that $\mathbf{E} \cdot \mathbf{n} = 0$ for the plane with the circle cut out because $\mathbf{E}$ is in the plane and $\mathbf{n}$ is normal to it. By symmetry, if we get I for the flux through the square face, we should get $I/2$ for each of the rectangular faces, giving us

$$-\frac{Q}{2\epsilon_0} + \frac{4I}{2} + I = 0$$

$$3I = \frac{Q}{2\epsilon_0}$$

$$I = \frac{Q}{6\epsilon_0}.$$

In retrospect, this answer is not surprising; since there are six sides, each should have the same flux since the source is placed symmetrically between the sides.

APPLICATION 16.2 EXERCISES

1. Derive the Divergence Theorem in spherical coordinates. [*Hint:* Start with the Cartesian form.]
2. Find the flux through the portion of the plane $x + y + z = 1$ in the first octant due to a point charge Q at the origin.
3. The field produced by a dipole P along the z-axis, located at the origin, is given by

$$\mathbf{E} = \frac{2P\cos\phi}{r^3}\,\hat{\mathbf{r}} + \frac{P\sin\phi}{r^3}\,\boldsymbol{\phi}.$$

Find the flux through a disk of radius 1 unit, parallel the xy-plane, 1 unit away from the dipole along the z-axis. (See the figure.)

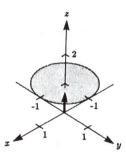

FIGURE FOR EXERCISE 3

4. For a harmonic function, $\nabla^2 u = 0$. So we know that $\iint_R \nabla^2 u\,dA = 0$, but $\nabla^2 u = \nabla \cdot \nabla u$. Therefore, we can use Green's Theorem to convert surface integrals to line integrals in the plane. Green's Theorem is, in vector form,

$$\iint_R \nabla \cdot \mathbf{V}\,dA = \int_C \mathbf{V} \cdot \mathbf{n}\,ds$$

where C is the boundary of R. So, by letting $\nabla u = \mathbf{V}$, we get

$$\iint_R \nabla \cdot \nabla u\,dA = \int_C \nabla u \cdot \mathbf{n}\,ds = \int_C \frac{\partial u}{\partial n}\,ds = 0.$$

The last integral states that the net flow out of the region where the solution exists must be zero. It can also be interpreted to mean that the average flow is zero. In light of this result, we see that solutions to $\nabla^2 u = 0$ have restrictions as to what is an acceptable physical boundary condition. Give a physical interpretation to $\int_C \nabla u \cdot \mathbf{n}\,ds = 0$.

5. Show that the function $\ln(x^2 + y^2)$ is harmonic in any region that excludes the origin.
6. Integrate the normal derivative of the function in Exercise 5 around a unit square with corners at $(1, 1)$, $(1, 2)$, $(2, 2)$, and $(2, 1)$.

APPLICATION 16.3: Modifying Ampere's Law:
When Current Changes With Time

> **See Section 16.7**
> **Stoke's Theorem**
> **Calculus, 4th Edition,**
> **Larson/Hostetler/Edwards**

In magnetostatics, we often use Ampere's Law. The integral form of this law states that the magnetic field **B** integrated around a path depends only on the enclosed current I, or

$$\int_C \mathbf{B} \cdot d\mathbf{l} = \mu_0 I$$

where μ_0 is the permeability of free space, and is constant. The differential form is

$$\nabla \times \mathbf{B} = \mu_0 \mathbf{J}$$

where **J** is the current density. We can see that the two forms are equivalent by applying Stokes's Theorem. We then get

$$\int\int_R \nabla \times \mathbf{B}\, dA = \int\int_R \mu_0 \mathbf{J} \cdot \mathbf{n}\, dA$$
$$= u_0 I$$
$$= \int_C \mathbf{B} \cdot d\mathbf{l}$$

where the last line is by Stokes's Theorem.

A simple illustration can show us that time-independent Ampere's Law does not work if the current is changing with time. This is the same correction Maxwell made to Ampere's Equation when he included it in his now-famous Maxwell's Equations for electrodynamics. Consider Figure 16.5 which represents a parallel-plate capacitor with wire leads to it.

Parallel Plate Capacitor

FIGURE 16.5

Assume a current is permitted to begin flowing in the wire toward one of the parallel plates. This will cause excess charge to build up on one plate, producing an electric field which will attract the opposite charge to the other plate (or equivalent). This process will continue until the electric field between the parallel plates is large enough to prevent additional charges to be deposited on the plates.

We will assume that we have not reached the steady state situation and that the current is still flowing at a changing rate. We also know that the current is $I = dQ/dt$. Applying Ampere's Law, we get

$$\int_C \mathbf{B} \cdot d\mathbf{l} = \mu_0 \int_R \int \mathbf{J} \cdot \mathbf{n} \, dA = \mu_0 I.$$

By taking a circular path around the conducting wire (but not too close to the capacitor), we simply get $2\pi r B = \mu_0 I$, or $B = \mu_0 I / 2\pi r$, where $\mathbf{B}$ is tangent to the circular path. This is the magnetic field associated with a long wire carrying current I.

EXAMPLE

The Result of Changing Current

Suppose we have the same parallel plate capacitor and the surface, as indicated in Figure 16.6. What is the result of changing current in this case?

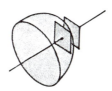

FIGURE 16.6

SOLUTION

The integral involving $\mathbf{B}$ is the same since the bordering curve is circular, as before. However,

$$\mu_0 \int_R \int \mathbf{J} \cdot \mathbf{n} \, dA = 0$$

since there is no current density passing through this surface. But there is an electric field $\mathbf{E}$ between the parallel plates. In fact, the electric flux over this same surface is $\int\int_R \mathbf{E} \cdot n \, dA = \phi_E$. Since the charge on the plates of the capacitor is changing, we expect to see the electric flux change. The results for $\int_C \mathbf{B} \cdot d\mathbf{l}$ will then be independent of which surface we choose if we write

$$\int_C \mathbf{B} \cdot d\mathbf{l} = \mu_0 I + \mu_0 \epsilon_0 \frac{d\phi_s}{dt}$$

where ϵ_0 is the permittivity of free space. This result can be interpreted to mean that the changing current results in a changing electric flux. This changing electric flux looks essentially the same as a current; it is often called a displacement current and notated as I_D.

APPLICATION 16.3 EXERCISES

1. Show that the magnetic field current beyond the edges of the capacitor in the example is given by

$$B = \frac{\mu_0 I_D}{2\pi r}.$$

2. Assume that our parallel-plate capacitor is made of circular disks, with a diameter of 20 centimeters. Also suppose that the capacitor plates are 2 millimeters apart. The voltage applied to the capacitor is $V = 120\cos(120\pi t)$. (a) Find the maximum magnetic field just outside the conductors leading to the capacitor and (b) the maximum magnetic field inside the capacitor. Assume the wire lead diameter of 5mm and that the electric field between the parallel plates is uniform with no electric field outside the plates.

3. A long coaxial capacitor (see the figure) has a changing voltage applied to it. The inner conductor has a radius of 2 centimeters and the outer conductor has a radius of 3 centimeters. The applied voltage is $120\cos(120\pi t)$. Find the displacement current per unit length (per meter) between the conductors.

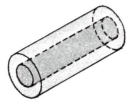

Coaxial Capacitor

FIGURE FOR EXERCISE 3

17

DIFFERENTIAL EQUATIONS

APPLICATION 17.1: Automobile Design: Drag Force

> **See Section 17.2**
> **Separation of Variables in First-Order Equations**
> **Calculus, 4th Edition,**
> **Larson/Hostetler/Edwards**

Contemporary automobiles are designed to meet consumers' demands for style and economy. A sleek powerful car may be the ideal for many, but high power may also bring high fuel costs. Newton's Law of Motion, including the "drag force," is

$$m\frac{dv}{dt} = F - R.$$

Here R is the retarding force, called the drag force, due in part to the cross-sectional area at the front of the car as it moves through the air. Specifically, for moderate to high velocities, R can be expressed as

$$R = \frac{1}{2}C\rho Av^2$$

where ρ is the density of air (0.0804 pounds/cubic foot or 0.00251 slugs/cubic foot), A is the frontal area of the car (typically 25 square feet), v is the speed, and C is the so-called *drag coefficient*, an experimentally obtained value. Notice that in this model, the drag force R is proportional to the square of the speed.

By substitution of values, the force needed for a car with a given drag coefficient can be determined. Then the corresponding horsepower that will be required can be quickly computed.

EXAMPLE

Comparing Drag Coefficient and Horsepower

Suppose two cars with typical cross-sectional areas are designed with drag coefficients of 0.5 and 0.25, respectively. Use Newton's Law of Motion and the relationship for drag force to find the difference in force necessary for each of the two cars to maintain a speed of 60 miles/hour (88 feet/second).

SOLUTION

We begin by considering Newton's Law of Motion.

$$m\frac{dv}{dt} = F - \frac{1}{2}C\rho Av^2$$

Since we are concerned with a constant speed, the acceleration is $dv/dt = 0$. When we substitute values ($A = 25$ ft^2, $\rho = 0.00251$ slugs/cubic foot, $v = 88$ feet/second) into the drag force equation, we get the force needed for each car. The force needed for the car with the 0.5 drag coefficient is 121.5 pounds, whereas the force needed for the car with the 0.25 drag coefficient is 60.75 pounds.

Simple physics tells us that the power being used is Fv, which gives, for one car, 10,692 foot-pounds/second and, for the other, 5346 foot-pounds/second. If we divide each by 550, we obtain the horsepower. The larger-drag-coefficient automobile will require 19.4 horsepower, whereas the lower-drag-efficient automobile will require 9.72 horsepower.

APPLICATION 17.1 EXERCISES

1. Ignoring all else, suppose a 3200 pound car (100 slugs) with drag coefficient of 0.5 has a terminal speed of 176 feet/second. Assume the force needed to get the car starting at rest to terminal speed is constant. Determine the time it will take for this car to accelerate from zero to 60 miles/hour (88 feet/second).

2. Assume the auto in Exercise 1 is redesigned so that the drag coefficient is reduced to 0.25. Find the new terminal velocity.

3. Assume the auto in Exercise 2 is to accelerate as did the auto in Exercise 1. How long will it take to get to 88 feet/second?

4. Full supertankers cruise at about 10 knots (a nautical mile is about 1.15 statute miles). This means that the engines can provide a force necessary to offset the drag provided by the hull in the ocean at 10 knots. Suppose a supertanker, cruising at 10 knots, reverses its propeller direction. It comes to rest in three hours at a position 15 nautical miles from the place at which it "applied the brakes." Calculate the drag force. Assume the force is proportional to v^2 and the mass is 5×10^8 kilograms.

5. For supertankers there is not much turbulence. Drag is more likely to be proportional to the speed rather than the speed squared. If this is the appropriate model, find the drag force that will stop a supertanker, cruising at 10 knots, in three hours at 15 nautical miles from the position at which it "applies its brakes." Assume the supertanker's mass is 5×10^8 kilograms.

APPLICATION 17.2: Coaxial Conductor

> See Section 17.2
> Separation of Variables in First-Order Equations
> Calculus, 4th Edition,
> Larson/Hostetler/Edwards

The coaxial conductors shown in Figure 17.1 has a constant voltage applied between the two conductors. Its potential function U satisfies Laplace's Equation. We can most easily use polar coordinates for the coaxial configuration and express Laplace's Equation as follows.

$$\nabla^2 U = 0$$

$$\frac{d^2 U}{dx^2} + \frac{d^2 U}{dy^2} = 0 \qquad \text{Rectangular form}$$

$$\frac{1}{r}\frac{\partial}{\partial r}\left(r\frac{\partial U}{\partial r}\right) + \frac{1}{r^2}\frac{\partial^2 U}{\partial \theta^2} = 0 \qquad \text{Polar form}$$

The potential U of the coaxial cable can thus be determined by solving Laplace's Equation.

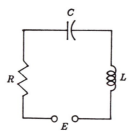

FIGURE 17.1

EXAMPLE

Finding Potential

Assuming a static situation within the conductor, find the potential U as a function of r for the coaxial cable described above. (See Figure 17.2.)

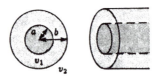

Coaxial Conductor

FIGURE 17.2

SOLUTION

Since we assumed a static situation within the conductor, the potential on each conductor will be constant. Therefore, we expect there will be no angular dependence for U between the conductors. The partial differential equation then reduces to the ordinary differential equation

$$\frac{1}{r}\frac{d}{dr}\left(r\frac{dU}{dr}\right)=0$$

$$r\frac{dU}{dr}=C_1$$

where C_1 is a constant. Then we have

$$\frac{dU}{dr}=\frac{C_1}{r}$$

$$dU=\frac{C_1}{r}\,dr$$

$$U=C_1\ln r+C_2$$

where C_2 is a constant. Since $U(a)=V_1=C_1\ln a+C_2$ and $U(b)=V_2=C_1\ln b+C_2$ we have

$$C_1=\frac{V_1-V_2}{\ln a-\ln b}\qquad\text{and}\qquad C_2=\frac{V_2\ln a-V_1\ln b}{\ln a-\ln b}$$

or collectively,

$$U(r)=\frac{V_1-V_2}{\ln a-\ln b}\ln r+\frac{V_2\ln a-V_1\ln b}{\ln a-\ln b}.$$

APPLICATION 17.2 EXERCISES

1. Show that Laplace's Equation $\nabla^2 U=0$ in polar coordinates is

$$\frac{1}{r}\frac{\partial}{\partial r}\left(r\frac{\partial U}{\partial r}\right)+\frac{1}{r^2}\frac{\partial^2 U}{\partial\theta^2}=0.$$

2. Find the potential between concentric conducting spheres of radii a and b if $U(a)=V_1$ and $U(b)=V_2$. *Hint:* The radial part of the Laplace operator is

$$\frac{1}{\rho^2}\frac{\partial}{\partial\rho}\left[\rho^2\frac{\partial U}{\partial\rho}\right].$$

3. If U is a function of three variables, then Laplace's Equation takes the form

$$\frac{d^2U}{dx^2}+\frac{d^2U}{dy^2}+\frac{d^2U}{dz^2}=0.$$

Show that the spherical coordinate form of this equation is

$$\frac{1}{\rho^2}\frac{\partial}{\partial\rho}\left[\rho^2\frac{\partial U}{\partial\rho}\right]+\frac{1}{\rho^2\sin\phi}\frac{\partial}{\partial\phi}\left(\sin\phi\frac{\partial U}{\partial\phi}\right)+\frac{1}{\rho^2\sin^2\phi}\frac{\partial^2U}{\partial\theta^2}.$$

4. Parabolic coordinates are related to spherical coordinates through the following transformation.

$$\rho=\frac{s}{1-\cos\phi},\qquad\rho=\frac{t}{1+\cos\phi},\qquad\theta=\theta$$

For s or t held constant, the resulting equation is a parabola involving variables ρ and ϕ. Obtain the Laplacian in parabolic coordinates.

APPLICATION 17.3: Oil Spill

> **See Section 17.2**
> **Separation of Variables in First-Order Equations**
> **Calculus, 4th Edition,**
> **Larson/Hostetler/Edwards**

We are all too familiar with the reported disaster: an oil spill from a ruptured tanker causes enormous damage to the ecology of an entire region. In the wake of the 1989 Alaskan oil spill, it became painfully obvious that existing methods were only partially successful in mitigating damages. Concerned citizens then called for renewed efforts in several directions—to minimize the chance of a recurrence; to develop more effective technology; and to respond faster in the event another spill occurs. For the latter, *time* is the crucial factor.

Let's consider a simple but related problem in which we determine the time it would take for a tank of liquid (water) to empty.

EXAMPLE

Emptying a Tank

We wish to determine how long it will take for a tank filled with water to empty. A cylindrical tank of radius 4 feet and height 12 feet is initially filled with water. A circular spigot of radius 1 inch is opened at the bottom of the tank and the water begins to drain out. (See Figure 17.3.)

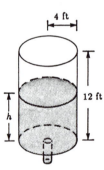

FIGURE 17.3

We assume that Torricelli's Law applies so that the velocity of the water through the spigot is given by

$$v = \sqrt{2gh}$$

where h is the height of the water. Determine the amount of time that it will take for the tank to empty completely.

SOLUTION

We begin by equating the decrease in the volume of water in the tank to the volume of water moving out through the spigot. The decrease in the volume of water in the tank is given by

$$(\text{area})(\text{height}) = -(16\pi)(\Delta h).$$

The volume of water moving out through the spigot is given by

$$v \, \Delta t \left(\frac{\pi}{12^2} \right) = \sqrt{2gh} \left(\frac{\pi}{144} \right) \, \Delta t.$$

Setting these equal, we have

$$-\pi 16\,\Delta h = \frac{\pi}{144}\sqrt{2gh}\,\Delta t$$

$$\frac{\Delta h}{\Delta t} = -\frac{1}{2304}\sqrt{2gh}.$$

Letting $t \to 0$, we obtain

$$\frac{dh}{dt} = -\frac{1}{2304}\sqrt{2gh}.$$

We solve this equation with initial condition $t = 0$ and $h = 12$. Separating variables produces

$$\frac{dh}{h^{1/2}} = -\frac{\sqrt{2g}}{2304}\,dt$$

$$2h^{1/2} = -\frac{\sqrt{2g}}{2304}t + C.$$

Substituting the initial condition yields $C = 2\sqrt{12}$. So,

$$2h^{1/2} = -\frac{\sqrt{2g}}{2304}t + 2\sqrt{12}.$$

Finally, the tank is empty when $h = 0$. We solve the resulting equation for t.

$$0 = -\frac{\sqrt{2g}}{2304}t + 2\sqrt{12}$$

$$t = \frac{2\sqrt{12}}{\sqrt{2g}}(2304) \approx 1995 \text{ seconds.}$$

APPLICATION 17.3 EXERCISES

1. Solve the example but assume $v = 0.6\sqrt{2gh}$.

2. Solve the example for a tank in the shape of a hemisphere of radius 4 feet. (See the figure.)

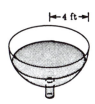

FIGURE FOR EXERCISE 2

3. Solve the example for a tank in the shape of a cone of radius 4 feet and height 12 feet. (See the figure.)

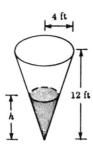

FIGURE FOR EXERCISE 3

4. Solve the example but assume that the cylinder is on its side. (See the figure.)

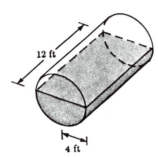

FIGURE FOR EXERCISE 4

APPLICATION 17.4: The Pendulum Revisited

> **See Section 17.2**
> **Separation of Variables in First-Order Equations**
> **Calculus, 4th Edition,**
> **Larson/Hostetler/Edwards**

Often, second-order nonlinear equations cannot be solved exactly. In some cases knowledge of the general behavior is all that is really needed to solve the problem.

Consider a pendulum of length L whose angular displacement θ is described by a nonlinear equation. We have

$$\frac{d^2\theta}{dt^2} + \frac{g}{L}\sin\theta = 0.$$

We begin by noting that the angular velocity ω is given by

$$\frac{d\theta}{dt} = \omega \quad \text{and} \quad \frac{d^2\theta}{dt^2} = \frac{d\omega}{dt}.$$

Then, we have

$$\frac{d^2\theta}{dt^2} = \frac{d\omega}{dt} = \frac{d\omega}{d\theta}\frac{d\theta}{dt} = \frac{d\omega}{d\theta}\omega$$

which implies that

$$\frac{\omega\,d\omega}{d\theta} + \frac{g}{L}\sin\theta = 0$$
$$\omega\,d\omega + \frac{g}{L}\sin\theta\,d\theta = 0.$$

Using separation of variables and integrating, we find the general solution to be

$$\frac{\omega^2}{2} - \frac{g}{L}\cos\theta = k.$$

We can obtain a graphic solution in the phase plane; this will consist of sketches of position versus velocity.

EXAMPLE

Phase Plane Techniques

In order to get a specific example for illustration, we will assume that in our pendulum problem $\theta = 0$ and $\omega = 1$ when $t = 0$. Solve the equation by a graph of the phase plane. Show how the solution depends on L.

SOLUTION

Substitution gives us

$$k = \frac{1}{2} - \frac{g}{L}$$

and

$$\omega = \pm\sqrt{1 - \frac{2g}{L} + \frac{2g}{L}\cos\theta}.$$

Now, depending on the size of g/L, or more specifically, on L since g is constant, we get the following curves.

For $2g/L = 1$, we get $\omega = \pm\sqrt{\cos\theta}$. (See Figure 17.4.)

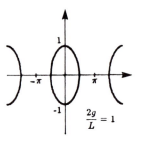

FIGURE 17.4

For $2g/L = 1/2$, we get $\omega = \pm\sqrt{\frac{1}{2} + \frac{1}{2}\cos\theta}$. (See Figure 17.5.)

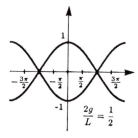

FIGURE 17.5

For $2g/L = 0.2$, we get open curves, as illustrated in Figure 17.6. This corresponds to a long pendulum with an initial linear velocity large enough so the pendulum does not swing back and forth, but instead loops completely around.

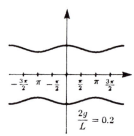

FIGURE 17.6

Recall that the initial conditions are $\theta = 0$ and $\omega = 1$. We see that as L becomes larger with the given initial conditions, the pendulum is able to swing completely around, whereas a small length L results in ordinary pendulum motion.

For an overview of several possible trajectories, see Figure 17.7. This figure illustrates values of $2g/L$ between 0.1 and 1.1 in steps of 0.2.

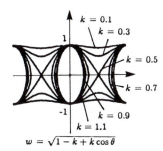

FIGURE 17.7

APPLICATION 17.4 EXERCISES

1. Show that the tangent is vertical for $2g/L > 1/2$ when $\omega = 0$. Explain the physical meaning.

2. Solve the linearized pendulum problem $y'' + \frac{g}{L}y = 0$ by the phase plane technique.

3. Springs do not necessarily obey Hooke's Law. A better model describing spring displacement is

$$\frac{d^2 x}{dt^2} + \alpha x + \beta x^3 = 0.$$

When $\beta > 0$, the spring is called a hard spring, and when $\beta < 0$, the spring is called a soft spring. Begin by letting $dx/dt = y$, we have $dy/dt = -\alpha x - \beta x^3$, which implies that

$$\frac{dy}{dx} = \frac{dy/dt}{dx/dt} = -\frac{\alpha y + \beta x^3}{y}.$$

Plot the trajectories for $\alpha = 1$ and $-0.5 \leq \beta \leq 0.5$. Assume $x = 0$ and $dx/dt = 1$ when $t = 1$.

4. The van der Pol Equation occurs as a result of a positive linear feedback and a negative cubic feedback. Electronic oscillators that use vacuum tubes or transistors can be described by a van der Pol Equation that reduces to

$$\frac{d^2 y}{dt^2} - \mu(1 - y^2)\frac{dy}{dt} + y = 0.$$

Substituting $v(dv/dy) = d^2y/dt^2$ and rewriting, we obtain $dv/dy = \text{constant} = k = f(v, y)$. (a) Plot the resulting curves $f(v, y) = k$ for $k = -5$, -1, 0, $1/4$, $1/2$ and 1, when $\mu = 0.1$, and (b) then construct a second plot with the same values of k with $\mu = 1$.

5. For the curves sketched in Exercise 4, draw small marks along each curve with slope equal to the particular k value and then sketch curves that smoothly fit from one slope to another. You will notice that in both cases ($\mu = 0.1$ and $\mu = 1$,) the curves approach a closed repeating curve as limits.

APPLICATION 17.5: Floating Buoy

> See Section 17.6
> Second-Order Nonhomogeneous Linear Equations
> Calculus, 4th Edition,
> Larson/Hostetler/Edwards

A cylindrical buoy has radius a, height h, and mass density ρ (less than the density of water). The buoy is released at the surface of the water. (See Figure 17.8.) By Archimedes' Principle, an object submerged in a fluid is buoyed up by a force equal to the weight of the displaced fluid.

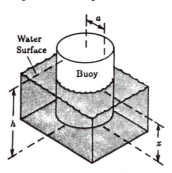

FIGURE 17.8

EXAMPLE

Motion of a Buoy

The buoy in Figure 17.8 is floating in water. Describe its motion.

SOLUTION

Let x represent the depth of the bottom of the buoy. Assume that the density of water is 1 gram per cubic centimeter. There are two forces: the gravitational force given by $mg = \rho \pi a^2 h g$ and the buoyancy given by $\pi a^2 x g$. Thus, we wish to solve the differential equation

$$m\frac{d^2 x}{dt^2} = mg - \pi a^2 x g$$

$$\frac{d^2 x}{dt^2} + \frac{\pi a^2 g}{m} x = g$$

subject to the initial conditions $t = 0$, $x = 0$, and $dx/dt = 0$. We begin by solving the homogeneous equation

$$\frac{d^2 x}{dt^2} + \frac{\pi a^2 g}{m} x = 0.$$

Since the characteristic equation has complex roots

$$\pm \sqrt{\frac{\pi a^2 g}{m}}\, i = \pm \sqrt{\frac{\pi a^2 g}{\rho \pi a^2 h}}\, i = \pm \sqrt{\frac{g}{\rho h}}\, i$$

we find the solution to be

$$x_h = C_1 \cos \sqrt{\frac{g}{\rho h}}\, t + C_2 \sin \sqrt{\frac{g}{\rho h}}\, t.$$

We next use the method of undetermined coefficients to determine that a particular solution of

$$\frac{d^2x}{dt^2} + \frac{\pi a^2 g}{m} x = g$$

is $x_p = \rho h$. Thus, the general solution is

$$x = x_h + x_p = C_1 \cos \sqrt{\frac{g}{\rho h}}\, t + C_2 \sin \sqrt{\frac{g}{\rho h}}\, t + \rho h.$$

We next apply the initial conditions. Setting $t = 0$ and $x = 0$, we obtain $0 = C_1 + \rho h$ and thus, $C_1 = -\rho h$. We next set $t = 0$ and $dx/dt = 0$.

$$\frac{dx}{dt} = -C_1 \sqrt{\frac{g}{\rho h}}\, \sin \sqrt{\frac{g}{\rho h}}\, t + C_2 \sqrt{\frac{g}{\rho h}}\, \cos \sqrt{\frac{g}{\rho h}}\, t$$

$$0 = C_2 \sqrt{\frac{g}{\rho h}}$$

So, $C_2 = 0$. Hence, the solution is

$$x = -\rho h \cos \sqrt{\frac{g}{\rho h}}\, t + \rho h.$$

Thus, the object oscillates about the position $x = \rho h$. Note that this is the equilibrium position.

APPLICATION 17.5 EXERCISES

1. Obtain the period, frequency, and amplitude of the motion.

2. If $a = 5$ centimeters, $h = 10$ centimeters, and the period 1 second, determine the mass of the buoy.

3. Work the example except assume that the buoy is a cube of sides 10 feet. Assume that the weight density of water is 62.5 pounds/cubic foot.

4. Work the example except assume that the buoy is a rectangular block of length 20 centimeters, width 10 centimeters, and height 5 centimeters.

5. Work the example except assume that the buoy is a sphere of radius 5 centimeters. Set up the differential equation only; do not solve the equation.

APPLICATION 17.6: Deflection of a Beam

> **See Section 17.6**
> **Second-Order Nonhomogeneous Linear Equations**
> **Calculus, 4th Edition,**
> **Larson/Hostetler/Edwards**

Bookshelves have a tendency to sag in the middle. Through a consideration of the initial conditions along with solving a fourth-order differential equation, we can find the shape of the curve that describes this sag. We assume the formula

$$EI\,\frac{d^4y}{dx^4} = f(x).$$

This is a reasonable approximation if $(dy/dx)^2$ is much smaller than 1, where E is Young's Modulus of Elasticity and I is the moment of inertia of the cross section of the shelf with respect to a horizontal line through the center of mass of the cross section. For most bookcases, this sagging is gradual and thus the slope, dy/dx, would be small. $f(x)$ represents the density of the vertical force at a point x.

EXAMPLE

The Sagging Bookshelf

We assume that the shelf has length L and constant weight density $f(x) = w$ and that the ends are secured so that at each end $y = 0$ and $y' = 0$. (See Figure 17.9.) What is the shape of the curve y that describes the sag in terms of x, E, and I? (Physically, $y = 0$ means the end is fixed, $y' = 0$ means the slope is zero, $y'' = 0$ means the curvature is zero, and $y''' = 0$ means there is no shear.)

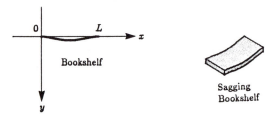

Bookshelf

Sagging Bookshelf

FIGURE 17.9

SOLUTION

For convenience we will show y in the downward direction as positive. By assumption,

$$\frac{d^4y}{dx^4} = \frac{w}{EI}.$$

We begin by integrating four times to obtain

$$y = \frac{1}{EI}\left(\frac{1}{24}wx^4 + C_1x^3 + C_2x^2 + C_3x + C_4\right).$$

We first apply the initial conditions $x = 0$, $y = 0$ and $y' = 0$ to obtain $C_3 = 0$ and $C_4 = 0$. Next we apply the initial conditions $x = L$, $y = 0$ and $y' = 0$ to obtain

and

$$0 = \frac{1}{6}wL^3 + 3C_1 L^2 + 2C_2 L.$$

We multiply the first equation by 2 and the second by L and subtract to get

$$C_1 = -\frac{1}{12}wL.$$

Substituting this into the first equation, we obtain

$$C_2 = \frac{1}{24}wL.$$

Thus,

$$y = \frac{1}{EI}\left(\frac{1}{24}wx^4 - \frac{1}{12}wLx^3 + \frac{1}{24}wL^2 x^2\right).$$

APPLICATION 17.6 EXERCISES

1. Determine the maximum value for y (the maximum sag).

2. Solve the example with the initial conditions $y(0) = y(L) = 0$ and $y''(0) = y''(L) = 0$. Find the maximum value for y.

3. Solve the example with the initial conditions $y(0) = y'(0) = 0$ and $y''(L) = y'''(L) = 0$. Find the maximum value for y.

4. Solve the example with the initial conditions $y(0) = y'(0) = 0$ and $y(L) = y''(L) = 0$. Find the maximum value for y.

APPLICATION 17.7: Electrical Circuit Differential Equation

> See Section 17.6
> Second-Order Nonhomogeneous Linear Equations
> Calculus, 4th Edition,
> Larson/Hostetler/Edwards

An RLC electrical circuit has inductance L (in henrys), resistance R (in ohms), capacitance C (in farads), and electromotive force E (in volts). (See Figure 17.10.) The differential equation for the circuit may be stated as

$$\frac{d^2q}{dt^2} + \left(\frac{R}{L}\right)\frac{dq}{dt} + \left(\frac{1}{LC}\right)q = \left(\frac{1}{L}\right)E(t)$$

where q is the charge of the capacitor in coulombs. This differential equation can be solved by use of a Laplace Transform.

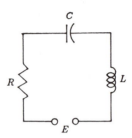

FIGURE 17.10

EXAMPLE

Using the Laplace Transform

An LC circuit has inductance $L = 10^{-3}$ henrys, capacitance $C = 10^{-9}$ farads, and constant voltage source $E = 12$ volts. We assume that the resistance R is zero. The initial conditions are $q(0) = q'(0) = 0$. Thus, we can assume the differential equation

$$q''(t) + \frac{1}{LC}q(t) = \frac{E}{L}.$$

The Laplace Transform of a function $f(t)$ is defined as

$$\mathcal{L}[f(t)] = \int_0^\infty e^{-st}f(t)\,dt.$$

Use the Laplace Transform to solve for the charge q as a function of time, subject to the initial conditions.

SOLUTION

For $L = 10^{-3}$, $C = 10^{-9}$, and $E = 12$, the differential equation becomes

$$q''(t) + 10^{12}q(t) = 12,000.$$

We shall apply the Laplace Transform to this differential equation using the initial conditions $q(0) = q'(0) = 0$. Using integration by parts, we obtain

$$\mathcal{L}[q'(t)] = \int_0^\infty e^{-st} q'(t)\, dt$$

$$= e^{-st} q(t) \Big]_0^\infty + s \int_0^\infty e^{-st} q(t)\, dt$$

$$= s\mathcal{L}[q(t)] - q(0)$$

where $\lim_{t \to \infty} e^{-st} q(t) = 0$. Using a similar argument, we obtain

$$\mathcal{L}[q''(t)] = s^2 \mathcal{L}[q(t)] - sq(0) - q'(0).$$

Since $q(0) = q'(0) = 0$, we have

$$\mathcal{L}[q''(t)] = s^2 \mathcal{L}[q(t)]$$

$$\mathcal{L}[12,000] = \int_0^\infty 12,000 e^{-st}\, dt$$

$$= -\frac{12,000}{s} e^{-st} \Big]_0^\infty = \frac{12,000}{s}, \quad \text{if } s > 0.$$

We can use the fact that

$$\mathcal{L}[af(t) + bg(t)] = a\mathcal{L}[f(t)] + b\mathcal{L}[g(t)]$$

where a and b are constants. This will enable us to split the differential equation as a sum of terms that are easier to deal with. (For a verification of this property, see Exercise 5.) When we apply Laplace Transform to the differential equation, we obtain

$$s^2 \mathcal{L}[q(t)] + 10^{12} \mathcal{L}[q(t)] = \frac{12,000}{s}.$$

So,

$$\mathcal{L}[q(t)] = \frac{12,000}{s(s^2 + 10^{12})}.$$

Using the method of partial fractions, we have

$$\mathcal{L}[q(t)] = (1.2 \times 10^{-8}) \left(\frac{1}{s} - \frac{s}{s^2 + 10^{12}} \right).$$

Since $\mathcal{L}[\cos(10^6 t)] = s/(s^2 + 10^{12})$. (See Exercise 1.) We obtain

$$q(t) = (1.2 \times 10^{-8})[1 - \cos(10^6 t)].$$

Thus, the solution of the differential equation may be obtained by applying the properties of the Laplace Transform.

APPLICATION 17.7 EXERCISES

1. Use the definition of the Laplace Transform to show that

$$\mathcal{L}[\cos(at)] = \frac{s}{s^2 + a^2}.$$

2. Use the definition of the Laplace Transform to show that

$$\mathcal{L}[\sin(at)] = \frac{a}{s^2 + a^2}.$$

3. Let $f(t) = t$. Find $\mathcal{L}[f(t)]$.

4. Let $f(t) = \begin{cases} 0, & 0 \le t \le 1 \\ 1, & 1 < t < 2 \\ 0, & 2 \le t. \end{cases}$

Find $\mathcal{L}[f(t)]$.

5. Show that the Laplace Transform has the property $\mathcal{L}[af(t) + bg(t)] = a\mathcal{L}[f(t)] + b\mathcal{L}[g(t)]$.

6. Show that $\mathcal{L}[f''(t)] = s^2\mathcal{L}[f(t)] - sf(0) - f'(0)$.

7. Solve the example except assume that the voltage is $E = \sin(120t)$.

8. Solve the example except assume that $L = 1$, $C = 10^{-2}$, and $E = \cos(50t)$.

APPLICATION 1.1

1. All the points of the circle given by $9000 = (x + 90)^2 + y^2$

APPLICATION 1.2

1. 38.97 ft³   2. 0.498 ft   3. 22.62 ft³   4. 14.14 ft³

APPLICATION 1.3

1. 9 hours   3. 1.2 Atm   4. 1.1 Atm

APPLICATION 1.4

1. $K = \frac{5}{9} F + 255\frac{2}{9}$ 2. $-40°$ F $= -40°$ C   3. 5798 K, 9977° F   4. 2700° F

APPLICATION 1.5

3. $15,000V$

APPLICATION 1.6

3. 1989 allocation: 80% of 45 = 36 units
   1989 charges: For usage of 36 units or less, use $1.50x$, where $x =$ number of units. For usage of more than 36 units, see the table.

Units	36	37	38	39	40	41	42	43
Fee	54	55.50	80.50	105.50	155.50	205.50	280.50	355.50

Units	44	45	46	47	48	49	50	
Fee	455.50	555.50	680.50	805.50	955.50	1105.50	1280.50	

APPLICATION 2.1

2. Escape velocity for the Moon $\approx$ 1.47 mi/sec   3. Escape velocity for Mars $\approx$ 3.09 mi/sec

4. Escape velocity for Planet X $\approx$ 2.64 mi/sec. Planet X is smaller than Earth.

APPLICATION 2.2

1. $\omega = \sqrt{48} = 4\sqrt{3}$ rad/sec   2. $k = \frac{1}{2}$ lb/ft   3. $m = \frac{5}{4}$ slugs   4. $\omega = 2$ rad/sec

APPLICATION 3.1

1. The transmitter must be placed at least $\left(\frac{2}{\sqrt{3}} - 1\right)$ units from the base of the hill.

2. $\sqrt{3}$ 3. Place the transmitter to the right of $(1, 0)$.

4. Place the transmitter at least $\frac{2}{15}$ units from the base of the hill.

APPLICATION 3.2

1. (a) rebound height = 1 ft   **(b)** velocity $v = \begin{cases} -32t, & 0 \le t < 1 \\ 8 - 32(t-1), & 1 < t < \frac{3}{2} \end{cases}$

2. velocity $v = \begin{cases} -32t, & 0 \le t < 1 \\ 4 - 32(t-1), & 1 < t < 1.25 \end{cases}$

3. (a) velocity $v = \begin{cases} -32t, & 0 \le t < 1 \\ 16\sqrt{3} - 32(t-1), & 1 < t < 1 + \sqrt{3} \end{cases}$

  (b) $\lim\limits_{t \to 1^-} v(t) = -32$

    $\lim\limits_{t \to 1^+} v(t) = 16\sqrt{3}$

  (c) No

4. velocity $v = \begin{cases} -32t, & 0 \le t < 1 \\ 16 - 32(t-1), & 1 < t < 2 \\ 8 - 32(t-2), & 2 < t < 2.5 \end{cases}$

APPLICATION 3.3

1. $\dfrac{dF}{dt} \approx 0.00003$ lb/sec   **2.** $\dfrac{dF}{dt} \approx -5.5$ lb/hr

3. $\dfrac{dr}{dt} \approx 2133.8$ mi/hr   **4.** weight ≈ 20.4 lb

APPLICATION 3.4

1. $\dfrac{dz}{dt} = \begin{cases} -\frac{5}{9}, & 0 \le x < 80 \\ -\frac{5}{4}, & 80 < x \end{cases}$

2. $\dfrac{dy}{dt} = -\dfrac{3600}{(120-x)^2}$ ft/sec, where x represents the distance from the man to the wall and $0 \le x \le 100$.

3. $\dfrac{dy}{dt} = \begin{cases} -\frac{75}{13}, & 0 \le x < 120 \\ -\frac{25}{4}, & 120 < x \end{cases}$ **4.** $\dfrac{dy}{dt} = \begin{cases} -\frac{10}{13}, & 0 \le x < 120 \\ -\frac{5}{4}, & 120 < x \end{cases}$

APPLICATION 3.5

1. (a) $V = \dfrac{4}{3}\pi \left\{ \left[\left(\dfrac{48}{\pi}\right)^{1/3} - \left(\dfrac{375}{4\pi}\right)^{1/3} \right] t + \left(\dfrac{375}{4\pi}\right)^{1/3} \right\}^3$ **(b)** About 5 hours

2. (a) $V = \left[-\dfrac{1}{3}t + 5 \right]^3$ **(b)** 15 hours

3. $V = \left(\dfrac{25}{6}\right)^3$ **4.** $V = \dfrac{1}{3}\pi(-2t + 3)^3$

APPLICATION 4.1

1. The light should be placed 9.2 ft above the floor.

2. The light should be place 10.6 ft above the floor.

3. $\tan\theta = \dfrac{8}{13}$ **4.** 16 ft^2

APPLICATION 4.2

1. $\phi = \dfrac{\pi}{4} + \dfrac{\theta}{2}$ **3.** $\theta \approx 32.98°$ **4.** Yes, 1.10

APPLICATION 4.4

1. $f = 20$ cm   **2.** (c) $v_0 = 200$ cm/sec   **3.** $k = 0.00111$

APPLICATION 4.5

2. The frequency changes most rapidly directly in front of (closest approach to) the source. The maximum change is $\dfrac{v_0{}^2 f}{cb}$.

3. $\dfrac{v_0{}^2 f}{cb} \approx 30.9 \approx 31$ HZ/sec

APPLICATION 4.6

2. $\delta = 51°$ **3.** $a = 1.348$, $b = 1.364$, $c = 1.400$ **4.** $\delta = 65°$

APPLICATION 4.7

1. $\Delta m \approx 0.001 m_0$ **2.** $v \approx 0.94c$

3. $\Delta E \approx 3.0325 \times 10^{-15}$ km m^2/sec$^2 = 3.0325 \times 10^{-15}$ joules

4. $\Delta E \approx 5.6 \times 10^{-12}$ km m^2/sec$^2 = 5.6 \times 10^{-12}$ joules

APPLICATION 4.8

1. (a) $\omega = \sqrt{\dfrac{PA}{\rho LV}}$, $f = \dfrac{\omega}{2\pi}$ (b) Yes

2. (b) $\Delta\omega = \dfrac{\omega_0}{30}$ (c) Yes, increases about 3%.

3. (a) $y_1 + y_2 = A\sin(kx + ct + \phi) + A\sin(kx - ct + \phi)$, $\sin(\alpha + \beta) + \sin(\alpha - \beta) = 2\sin\alpha\cos\beta$
Assuming that $\alpha = kx + \phi$, $\beta = ct$, we have $y_1 + y_2 = 2\sin(kx + \phi)\cos ct$.
The result is a standing wave.

(b) If $\phi = \pi/2$, $y_1 + y_2 = 2\cos kx\cos ct$, $\cos 0 = 1$, the result is an antinode.

(c) $\cos kx = 1$ or -1

$$kL = n\pi$$

$$k = \frac{n\pi}{L}$$

Fundamental $\lambda = 2m$ (half wave)

$$\lambda f = c$$

$$f_0 = \frac{L}{\lambda} = \frac{343}{2} = 171.5 \text{ hz}$$

$$f_1 = 2(171.5) = 343 \text{ hz}$$

$$f_2 = 3(171.5) = 514.5 \text{ hz}$$

$$f_n = n(171.5) \text{ hz}$$

(d) There is no displacement.

(e) We want a node, so the length must be $\frac{1}{2}$. This produces the series f, $3f_1$, $5f_1$, etc.

APPLICATION 5.1

1.

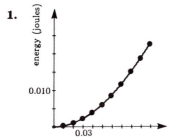

displacement (m)

2. $(0.02, 0.001)$, $(0.04, 0.004)$, $(0.06, 0.0086)$, $(0.08, 0.0150)$, $(0.10, 0.0226)$

3.

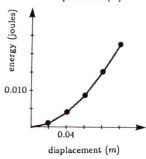

displacement (m)

4. $k \approx 5$

5. Use the Trapezoidal Rule to obtain the following data.

x	0.1	0.2	0.3	0.4	0.5	0.6	0.7	0.8	0.9	1.0
Energy	0.024	0.099	0.223	0.397	0.619	0.891	1.210	1.579	1.995	2.459

APPLICATION 5.2

2. $a_0 \approx 70.4$, $a_1 \approx 8.5$, $b_1 \approx 2.0$ **4.** $a_2 \approx -0.08$, $b_2 \approx 0.72$

5. $a_0 \approx 43$, $a_1 \approx -20.3$, $b_1 \approx -4.04$

APPLICATION 6.1

1. 211 lb   **2.** He should lose 2 lb.   **4.** She would lose 3.5 lb/day.

APPLICATION 6.2

1. 0.144 ft^2 **2.** approximately 66 mg   **3.** approximately 12 ml

APPLICATION 6.3

3. 0.124

APPLICATION 6.4

1. 2.5   **2.** $\log_{10} 3 \approx 0.48$ **3.** 5.8   **4.** $10^{1/2} \approx 3.16$

APPLICATION 6.5

1. 80.5 dB   **2.** 54.6 dB   **3.** infinite
4. 100 ft if the intensity drops off as $1/\text{distance}^2$.

APPLICATION 6.6

1. $t \approx 1.8414$ sec   **2.** $t \approx 1.6024$ sec   **3.** $t \approx 67.5$ sec   **4.** $t \approx 51.7875$ sec

APPLICATION 6.7

1. $t \approx 29$ min   **2.** (a) $x = 300 - 280e^{-(1/25)t}$ (b) 300 lb
3. (a) $0.0005 - 0.0004e^{-(1/4)t}$ (b) $t \approx 1.15$ years   **4.** $\dfrac{1}{e}$

APPLICATION 6.8

1. $\frac{3}{2}kT$ **2.** $T_c = 0.0417 K m_0 c^2$
$$T_r = 0.0472 K m_0 c^2$$

APPLICATION 7.1

3. 73.5 J

APPLICATION 7.2

1. $\delta \approx 0.515\pi$

APPLICATION 7.3

3. 14.11 N

APPLICATION 7.4

2. $985430 \text{ gm} - \text{cm}^2$ **3.** 0.456

APPLICATION 7.5

1. 399

APPLICATION 7.6

3. 28.6 yd^3

APPLICATION 7.7

3. 2580.6 ft-lb

APPLICATION 7.8

1. Work = 16,625 ft-lb   **2.** Work = 17,250 ft-lb   **3.** Work = 16,875 ft-lb

4. Work = 20,437.5 ft-lb   **5.** Work $\approx$ 16,199 ft-lb

APPLICATION 7.9

1. Work $\approx$ 0.85 ft-lb   **2.** Work = $\frac{3}{4}$ ft-lb

3. Work $= -\displaystyle\int_1^3 x\sqrt{1 + \frac{9}{4}\,\frac{(x-1)^2}{[4-(x-1)^2]}}\,dx$

4. Work $= \displaystyle\int_0^{625/4} x\sqrt{1 + \left(\frac{625 - 8x}{625}\right)^2}\,dx$

APPLICATION 7.10

3. 22980.8 lb/ft^2

APPLICATION 7.12

3. 18.607 ft

APPLICATION 8.1

1. (a) $\frac{1}{2}v^2 + 100x^2 = 100$ (b) $x = \cos(10\sqrt{2}\,t)$

2. (a) $\frac{1}{2}v^2 + 100x^2 = 150$ (b) $x = \cos(10\sqrt{2}\,t) + \frac{1}{\sqrt{2}}\sin(10\sqrt{2}\,t)$

3. $x = 0$, $v = 0$. There will be no motion because the initial velocity, initial position, and external forces are all zero.

4. (a) $\frac{1}{2}v^2 + 100x^2 = \frac{1}{2}v_0{}^2 + 100x_0{}^2$ (b) $x = x_0\cos(10\sqrt{2}\,t) + \frac{v_0}{10\sqrt{2}}\sin(10\sqrt{2}\,t)$

5. $\theta = \frac{\pi}{12}\cos(4t)$ **6.** $\theta = \frac{\pi}{6}\cos(2\pi t)$

APPLICATION 8.2

3. $y = \sqrt{\dfrac{c}{\pi}}\, h^{1/4}$, c is a constant.

APPLICATION 8.3

1. L'Hôpital's Rule is not helpful in evaluating this limit.

2. $\lim\limits_{t\to\infty} v(t) = c$ **3.** $\lim\limits_{t\to\infty} v(t) = c\dfrac{k}{\sqrt{k^2 + 1000^2 c^2}}$ **4.** $\lim\limits_{t\to\infty} v(t) = c$

APPLICATION 8.4

3. 0

APPLICATION 8.5

3. $\dfrac{T}{\sqrt{3}}$

APPLICATION 8.6

1. $x = 10\ln\left|\dfrac{10}{y} + \dfrac{\sqrt{100-y^2}}{y}\right| - \sqrt{100-y^2}$ **2.** $y = \dfrac{x^{3/2}}{9} - 3x^{1/2} + 6$

3. $y = \dfrac{x^{3/2}}{6} - 2x^{1/2} + \dfrac{8}{3}$

APPLICATION 8.7

1. $t = \dfrac{(5000)^{3/2}}{\sqrt{2g}(4000)}\left[\cos^{-1}\sqrt{\dfrac{4}{5}} + \dfrac{2}{5}\right]$

2. $\displaystyle\int \dfrac{\sqrt{8000x}}{\sqrt{8000-x}}\,dx = -(8000)^{3/2}\left[\cos^{-1}\dfrac{\sqrt{x}}{\sqrt{8000}} + \dfrac{\sqrt{8000-x}}{\sqrt{8000}}\,\dfrac{\sqrt{x}}{\sqrt{8000}}\right] + C_2$

3. $t = \dfrac{(2080)^{3/2}}{\sqrt{2(0.17g)}(1080)}\left[\cos^{-1}\sqrt{\dfrac{1080}{2080}} + \dfrac{\sqrt{1000}\sqrt{1080}}{2080}\right]$

4. $t = \dfrac{(3100)^{3/2}}{\sqrt{2(0.38g)}(2100)}\left[\cos^{-1}\sqrt{\dfrac{2100}{3100}} + \dfrac{\sqrt{1000}\sqrt{2100}}{3100}\right]$

APPLICATION 8.8

1. $m(f)(s) = -\dfrac{2^s + 1}{s}$, if $s < 0$.   **2.** $m(f)(s) = -\dfrac{1}{s+1}$, if $s < -1$.

APPLICATION 9.1

3. Yes, the sequence is finite.

APPLICATION 9.2

3. $0.99970c$

APPLICATION 9.3

1. 4

APPLICATION 9.4

1. $\mathcal{L}(f) = \dfrac{1}{s}[1 - e^{-s} + e^{-2s} - e^{-3s} + \cdots]$ **2.** $\mathcal{L}(f) = \dfrac{1}{s^2 + 1}$ **3.** $\mathcal{L}(f) = \dfrac{s}{s^2 + 1}$

4. $\mathcal{L}(f) = \dfrac{a}{s^2 + a^2}$ **5.** $\mathcal{L}(f) = \dfrac{s}{s^2 + a^2}$ **6.** $\mathcal{L}(f) = \dfrac{e^{1-s} - 1}{(1 - e^{-s})(1 - s)}$

APPLICATION 9.5

3. Yes, the field looks like a dipole field.

$$E = 2\pi\lambda k \left[\frac{a - z}{\sqrt{(z - a)^2 + k^2}} + \frac{z + a}{\sqrt{(z + a)^2 + k^2}} \right]$$

APPLICATION 9.6

1. $\theta = \dfrac{\pi}{6} - \left[8\sin\left(\dfrac{\pi}{12}\right) \right] t^2 + \dfrac{8}{3}t^4 - \cdots$ **3.** $y = 1 + x - \dfrac{1}{2}x^2 - \dfrac{1}{3}x^3 + \dfrac{1}{8}x^4 + \cdots$

APPLICATION 10.1

1. $y = \dfrac{377}{(1750)^2}x^2$ **2.** length ≈ 3606 ft   **3.** $(y - c) = \dfrac{(500 - c)}{(1150)^2}x^2$

APPLICATION 11.1

3. The Moon always falls toward the Sun.

APPLICATION 11.2

3. 14.72

APPLICATION 12.1

3. 0.00375 nm

APPLICATION 12.3

3. $\dfrac{4\mu_0 I}{2\pi[x^2 + (L^2/64)]} \cdot \dfrac{L^2/64}{\sqrt{(L^2/32) + x^2}}$

APPLICATION 13.1

1. $\theta \approx 0.38$ radian, $v_0 = 200\sqrt{29}$ ft/sec   **2.** $\theta \approx 0.1$ radian, $v_0 = 200\sqrt{365}$ ft/sec

3. $\theta \approx 1.28$ radians, $t = 5.57$ sec

APPLICATION 13.3

3. $y = A + Bx + Cx^2 + Dx^3 + Ex^4 + Fx^5$

$A = 0.7$, $B = 0$, $C = -0.7143, D = -3.2137$, $E = 6.2431$, $F = -2.8738$

where x, y are in thousands of feet

APPLICATION 14.2

3. $I_c \approx 15.916$

$V_{ce} \approx 1.274918$

$I_b \approx 9.6746$

APPLICATION 14.3

1. $\dfrac{q}{4\pi\epsilon_0}\left[\dfrac{-10\,\mathbf{i}}{(125)^{3/2}} + \left(\dfrac{5}{(125)^{3/2}} - \dfrac{1}{25}\right)\mathbf{j}\right]$

3. $\dfrac{q}{4\pi\epsilon_0}\left[\left(\dfrac{(x - a)}{[(x - a)^2 + y^2]^{3/2}} - \dfrac{(x + a)}{[(x + a)^2 + y^2]^{3/2}}\right)\mathbf{i} + \left(\dfrac{y}{[(x - a)^2 + y^2]^{3/2}} - \dfrac{y}{[(x + a)^2 + y^2]^{3/2}}\right)\mathbf{j}\right]$

4. $-\dfrac{V_0}{ln(b/a)}\dfrac{1}{r}\overrightarrow{u_r}$ where $\overrightarrow{u_r}$ is the unit vector in the radial direction, i.e., $\overrightarrow{u_r} = \dfrac{x\mathbf{i} + y\mathbf{j}}{\sqrt{x^2 + y^2}}$

APPLICATION 14.4

2. $x \approx 10.6$ ft, $y \approx 10.6$ ft, $z \approx 8.9$ ft

3. Yes, if the heat loss through the ceiling is approximately 3.29 times as great as the heat loss through the floor.

APPLICATION 14.5

3.

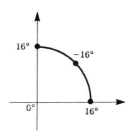

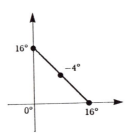

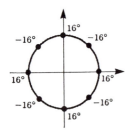

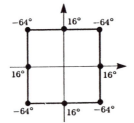

APPLICATION 15.1

1. 32.7 ft³/ hour   **2.** 2.04 hours/week   **3.** $6.154 \leq r \leq 7.154$ **4.** $0 \leq r \leq 1$

APPLICATION 15.2

3. The center of mass is $\frac{2}{9}$ m toward the sphere from the center of the rod.

APPLICATION 15.4

3. 462.5 m²

APPLICATION 15.5

1. $F_z = \dfrac{Gm\rho\pi a^2 h}{[b+(h/2)]^2}$ **2.** $F_z \approx \dfrac{\pi a^2 h\rho Gm}{(b^2+bh)}$

3. $F_z = Gm\rho(2\pi)[b - (h^2 + b^2)^{1/2} - a + (h^2 + a^2)^{1/2}]$

4. $F_z = Gm\rho(2\pi)[a + h - (a^2 + h^2)^{1/2}]$

APPLICATION 16.1

1. $25k\left(\dfrac{\pi}{3}\right)\left(\dfrac{\sqrt{2}}{2}\right)$ **2.** $50k\pi$ **3.** $50k\pi$

APPLICATION 16.2

3. $\dfrac{\pi P}{(2)^{3/2}}$

APPLICATION 16.3

3. $-\dfrac{\mu_0}{2}\ln\left(\dfrac{3}{2}\right)\cdot 120^2\sin(120\pi t)$

APPLICATION 17.1

2. 248.9 ft/sec   **3.** 18.92 sec

APPLICATION 17.3

1. 3325 sec   **2.** 538 sec   **3.** 399 sec   **4.** 2074 sec

APPLICATION 17.4

3.

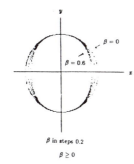

APPLICATION 17.5

1. Period $= 2\pi\sqrt{\dfrac{\rho h}{g}}$

Frequency $= \dfrac{1}{2\pi}\sqrt{\dfrac{g}{\rho h}}$

Amplitude $= \rho h$

2. 1948 gm   **3.** $x = -\dfrac{4}{25}\rho g\cos\dfrac{5}{2\sqrt{\rho}}t + \dfrac{4\rho}{25}g$ **4.** $x = -5\rho\cos\sqrt{\dfrac{g}{5\rho}}\,t + 5\rho$

5. $m\dfrac{d^2x}{dt^2} = mg - V(x)g$, where $m = \dfrac{500\pi}{3}\rho$ and $V(x) = \pi\left[\dfrac{15x^2 - x^3}{3}\right]$

APPLICATION 17.6

1. Maximum occurs at $x = \dfrac{L}{2}$.

2. $y = \dfrac{1}{EI}\left[\dfrac{1}{24}\omega x^4 - \dfrac{1}{12}\omega L x^3 + \dfrac{1}{24}\omega L^3 x\right]$ Maximum occurs at $x = \dfrac{L}{2}$.

3. $y = \dfrac{1}{EI}\left[\dfrac{1}{24}\omega x^4 - \dfrac{1}{6}\omega L x^3 + \dfrac{1}{4}\omega L^2 x^2\right]$ Maximum occurs at $x = L$.

4. $y = \dfrac{1}{EI}\left[\dfrac{1}{24}\omega x^4 - \dfrac{5}{48}\omega L x^3 + \dfrac{1}{16}\omega L^2 x^2\right]$ Maximum occurs at $x = \dfrac{15 - \sqrt{33}}{16}L$.

APPLICATION 17.7

3. $\dfrac{1}{s^2}$ **4.** $\dfrac{e^{-s} - e^{-2s}}{s}$

7. $q = \dfrac{1}{10^6} \dfrac{1}{(120^2 - 10^{12})} \sin(10^6 t) + \dfrac{1}{120(10^{12} - 120^2)} \sin(120t)$

8. $q = -\dfrac{1}{2400} \cos(50t) + \dfrac{1}{2400} \sin(10t)$